MANAGEMENT OF DESIGN
Engineering and Management Perspectives

MANAGEMENT OF DESIGN
Engineering and Management Perspectives

Edited by

Sriram Dasu
and
Charles Eastman

Kluwer Academic Publishers
Boston/Dordrecht/London

Distributors for North America:
Kluwer Academic Publishers
101 Philip Drive
Assinippi Park
Norwell, Massachusetts 02061 USA

Distributors for all other countries:
Kluwer Academic Publishers Group
Distribution Centre
Post Office Box 322
3300 AH Dordrecht, THE NETHERLANDS

Library of Congress Cataloging-in-Publication Data

Management of design / engineering and management perspectives / edited by Sriram Dasu and Charles Eastman.
p. cm.
Papers presented at a conference held at the University of California at Los Angeles in 1992.
Includes bibliographical references.
ISBN 0-7923-9509-3 (acid-free paper)
1. New products--United States--Management--Congresses. 2. Product management--United States--Congresses. 3. Design, Industrial--United States--Congresses. I. Dasu, Sriram. II. Eastman, Charles M.
HF5415.153.M333 1994
658.5'75--dc20 94-34778
CIP

Printed on acid-free paper.

Printed in the United States of America

CONTENTS

Preface

Design has always been one of the most challenging activities of mankind. We often judge a society by its products: the ancient Egyptians by their pyramids and temples, the Aztecs and Mayas by their temples and cities, the medieval Germans and Swiss by their precision instruments and watches. In the 1800s, the USA was renowned for the development of consumer products and mass production. Today, that reputation has moved to the Japanese, and in a faster evolution, to many developing countries, such as Korea and Taiwan. More and more, a country's balance of payments is determined by the value it adds to products and commodities. Some of the most intense value creating activities are the design and development of new products.

In an increasingly international marketplace the quality of the finished product is a major determinant of the success of a firm. Product design significantly influences product cost and quality, as well as market share and profitability of a firm. In many industries the prevailing competitive environment is such that firms have to offer a broad array of products that have to be frequently upgraded. The introduction of new products and their refinement, is an accelerating phenomenon. This is reflected in the variety of products available to customers. For instance, in 1992 a buyer of a printer for personnel computers had a choice of 126 printers produced by 41 companies. The price of the printers ranged from $350 to $10,000. Several firms offered 5 or more models[1]. The automobile industry is a good example of an industry in which there has been an increase in the rate at which new products are introduced. The number of passenger car lines sold in the US. has grown from 45 in 1971 to 65 in 1992 -- a 40% increase in two decades. In the 1950's the sales of a popular model, such as Chevy Impala, exceeded 1.5 million units. Today the sales of a single model rarely exceed 0.5 million. Product life cycles have also reduced. Most models are now upgraded every four years. A few decades ago no major changes were made for at least 8 years[2]. Similar changes can be observed in many other industries including semiconductors, consumer electronics, software, pharmaceuticals, and commercial aircraft.

Design is a creative activity that involves the synthesis of concepts and information drawn from different bodies of knowledge. As a result design projects often involve many people belonging to different functional areas such as various engineering disciplines, manufacturing, and marketing. In addition, in many organizations several design projects may be under way at the same

---

1 Byte Magazine, May 1992.

2 Wards Automotive World, 1971, 1991.

time. Due to this complexity, management of design has given rise to a rich set of research problems in management and engineering. Solution of these problems may have important impacts on new product costs and quality.

In this volume, we consider design as the planning and specification activity prior to fabrication. Design determines what products will be produced, how they will be produced, and when they will be introduced into the market. Enormous progress has been made in addressing the technical issues that arise at the detailed design phase. In comparison, considerably less attention has been paid to the overall organization of the development activities. It is now widely recognized that the quality of the products and the speed with which they are developed are significantly affected by the design process. The process by which a product is developed is determined by the managerial and engineering practices, tools and techniques.

This book is a collection of papers that present engineering and management perspectives on design. The issues studied can be classified into five broad topics:

(1) Decomposition of product development projects;
(2) Tools and techniques for preliminary evaluation of designs;
(3) Interface between design and manufacturing, assembly and distribution;
(4) Design information flows; and
(5) Determination of the scope, timing and duration of projects, and the allocation of resources.

The broad scope of the issues addressed by these papers reveals the current complexity of and sophistication in design management. Interestingly, the complexity in product design that we now observe is a recent phenomenon, arising largely in the last eighty years. Tracking the evolution of design management gives us a perspective on the complexities that now exist, in particular why and how they came about. Such a history, done thoroughly, would be a major undertaking in its own right. Here, in order to place the work of this volume in perspective, we give only a quick sketch, an overview that still awaits fleshing out.

The Changing Structure of Design

For most of history, design was an activity integral with the making of the object. The medieval draftsperson designed and fabricated utensils, early weapons, and clothes. Designs were transmitted between craftsmen in the form of the product itself. Even as late as the 14th century, for products as complex as clocks, there was no separation between design and fabrication; designs were generated by copying, either directly or from mental images, with layouts scribed directly on the material being worked. Only simple products were produced for trade. Complex products were produced "on consignment", with the owner agreeing to purchase the product before it was made. Thus initially, there was no clear distinction between design, fabrication and exchange.

The basic need for separating design arose first in architecture and civil works, which involved teams of people and took decades to complete. The earliest known design document is from Lagash Mesopotamia, 2900 BC.[3] and is a plan of a building, inscribed on a stone tablet. The Greeks and Romans both developed complex designs. The books of Vitruvius[4], written just before the time of Christ, remain as an important source of the early concepts and methods of architectural and civil design. Plans and elevations still exist that were used in the construction of medieval cathedrals[5].

Historical records indicate that throughout most of the medieval period design was not considered separate from construction. Architects were also foremen and heads of construction. Drawings were used by an architect to communicate to the building project overseers and to plan construction activities. Only later in the Renaissance did architecture separate itself from construction[6]. In shipbuilding also, there is clear historical evidence that design was separate from fabrication. Plans and section drawings exist of ancient ships[7].

The Industrial Revolution and the Emergence of the Design Function

The industrial revolution of the late 1700s resulted in a great proliferation of products, for both industrial and consumer use. Initially all products were custom made and hand fitted. In the early 1800s, with the harnessing of water and steam power to make shops and factories, the ability to fabricate products was much improved. These more complex tools required specialists to operate them. At the same time, the ideas of mass production, attributed to Eli Whitney for the production of arms, required that the designs

---

[3] Ernest de Sarzac, Decouvertes en Chaldee, 1891, Sterling Memorial Library, Yale University.

[4] Vitruvius, The Ten Books of Architecture, Trans. by M.H. Morgan, Dover Publications, New York 1914.

[5] Ackerman, J.S. "Ars Sine Scientia Nihil Est." Gothic Theory of Architecture at the Cathedral of Milan", Art Bulletin, June 1949, pp 84-111.

[6] Goldwaite, The Building of Renaissance Rome, The John Hopkins Press, Chapter 7.

[7] Booker, P.J, History of Engineering Drawing, Chatto and Windus, London 1963,

become explicit because each of the parts were made separately. It is interesting to note, that the first contracts for colt guns offered by the US. government were based on sample parts -- the product was the design[8].

Between 1800 and 1850, the manufacturing industries adapted architectural drafting to their needs, and drawings became the standard information source of a product design. Gaspard Monge's *La Geometrie descriptive*, published in 1801, formalized engineering drawing. The three orthogonal views of an object --the plan, elevation, and profile--that have been the basis of engineering drawing for the previous two centuries were formally presented for the first time by Monge. In the mass production environment preliminary analysis of the design became an economic necessity. Some early examples of design analysis are found in the drawings of armaments and forts. These works refer to maximum projectile paths and other military planning concepts that were used to both generate and evaluate designs[9].

Early in the nineteenth century engineering departments became separate parts of organizations. By 1850 not only was there a separation between design and fabrication but the design function itself was further partitioned into specializations. Draftsmen were recognized as a separate group that supported engineers and drafting was further specialized into drafters, tracers and checkers[10].

The Beginning of Design Management.

Around 1910, influenced by the studies of Frederick Taylor, manufacturing systems were reorganized. We observe the emergence of job specifications, time and motion studies and assembly lines. Procedures for manufacturing and assembling products were defined in great detail by the industrial engineers, allowing faster assembly and the production activity to be accomplished by lower skilled employees. The implication of industrial engineering concepts for design offices was that designers also had to pay greater attention to how the parts were being produced. The gains in productivity in the manufacturing operations suggested the need for an

---

[8] A. Pulos, American Design Ethic: History of Industrial Design to 1940, MIT Press, 1983.

[9] Diderot, Denis and d'Alembert, Encyclopedie ou Dictionaire raisonne des sciences, des arts et des metiers, 35 Volumes, Paris, 1751-1780; also G. Monge, Geometrie descriptive, Paris, 7th ed., 1847.

[10] O. Byrne, The American Engineer, Draftsman and Machinist's Assistant, C.A. Brown and Company, Philadelphia, 1853. Hemphill, J.C, "Systems in the drafting room", Machinery, November 1986.

analogous structuring of design operations, thus giving rise to design management.

To enhance design productivity, attempts were made to streamline the flow of information by borrowing ideas from manufacturing systems. The physical world of mechanical products was decomposed into discrete units whose fabrication could be broken into linear steps, in an assembly line. It was believed that the structure of a design could follow a similar decomposition. Articles on drafting room practice refer to the solution of "serializing the process" determining a standard flow through which all design work would pass from department to department[11]. An example of such fixed flows is shown in Figure 1, from a 1932 paper presenting machine design management at Western Electric[12].

Observe that in Figure 1 five different departments were involved in developing and approving the design. The figure not only reveals the existence of different engineering specialists, but also how the authority to initiate and approve changes was shared by different departments. Thus even with serial flow of design, in large organizations there was a need for managing the design process. Technical journals, beginning about 1890, mention some of the managerial issues, with fairly extensive discussions by the 1930s [13].

Early techniques for controlling large projects involved the use of department log sheets to record expected completion dates of tasks. Completion times were estimated from previous, similar projects. Log sheets were updated periodically, allowing problematic tasks to be identified -- those which could not be completed in the original time allocated and/or which would result in a delay of other project tasks. These project control charts can be viewed as the predecessors to modern tools such as PERT/CPM.

Associated with the project control charts were accounting systems which were used to monitor the accumulation of costs over the life of the project. Projects were decomposed into component activities; manpower and material costs were estimated for each. Tasks were given unique control numbers. Project members would then charge the cost of their time to these control numbers. Periodically, expenditure against control numbers could be

---

[11] W.B. Morton, "Proper procedure expedites design work", Electrical World, December, 1930.

[12] J.L. Alden, "Machine Design Management", Proc. Soc. of Mechanical Engineers, SMP-54-9, 1932.

[13] H.W. Alden, "The decimal index in the drawing room", Amer. Machinist, Feb 1897; G.F. Hamilton, "Simple drafting room methods", Industrial Management, October 1913.

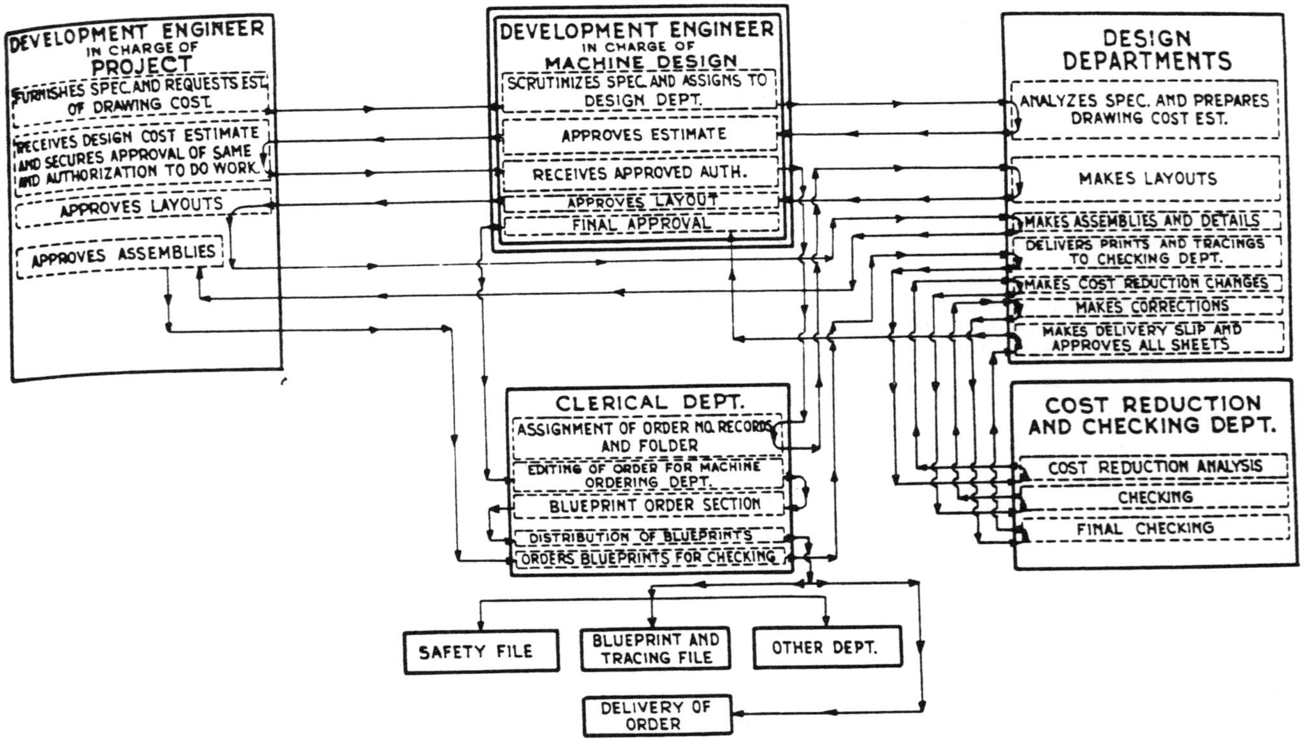

Figure 1: Fixed flow of information in machine design in Western Electric[28].

assessed relative to the pre-established budget. This approach remains in place in many design organizations today.

Concurrently with the huge growth in manufacturing, there was the expansion of our physical understanding of product behavior. This began with such simple things as cannon trajectories and continued later with the strength of homogeneous materials such as iron and steel. Engineers adopted the knowledge of physics and combined it with empirical experiments to develop much better predictions about the performance of designs before they were built. In this country, these capabilities began with the engineering of railway bridges[14]. The analysis quickly expanded to many forms of complex structures, laminar flow of fluids, and electrical flows. By the 1940s, this aspect of engineering became dominant and led to great increases in predictive modeling[15].

Each form of predictive modeling required its own representation of the product or part. These ranged from a few algebraic equations and the parameters for characterizing the specific design to elaborate simulation models such as scale models for use in wind tunnels. However, there were serious limitations in the application of predictive models to design, due to the limited amount of computation that could be done manually. Approximation techniques were very important. In many areas complexity did not allow computational techniques to be applied and mock-ups and the testing of scale models were important approaches in engineering research and practice. For example, the structural analysis of airplane wings was not possible until the advent of computers[16]. Even now, wind tunnel testing of scale models is utilized.

During the second world war significant advances were made in engineering principles and techniques for managing design and development. Many of the problems in management of design that modern corporations are facing today were also faced by the developers of weapon systems. This is best illustrated by the changes in the aircraft industry. In 1930s and 1940s, British and US. aircraft developers had to rapidly develop and produce many new aircraft. This urgency led to several innovations. Prior to the war, due to the high technical risk, the norm in the aircraft industry was to develop the product in a sequential manner and the different phases in the development were not overlapped. Also, the development of a new aircraft was almost entirely done

---

[14] H. Cowan, Science and Building, Wiley Interscience, New York, 1978.

[15] H.A. Simon, The Science of the Artificial, MIT Press, 1969, Chapter 3.

[16] Thomas Faulconer, Introduction to Aircraft Design, McGraw Hill, N.Y. 1942

by one firm[17]. To compress the development cycle several changes had to made. The design of the aircraft was partitioned into several components that were developed in parallel, often by different firms. Previously, production orders for new aircraft were released only after the prototype was approved. By 1940 it was quite common for several production steps to be initiated prior to the completion of the design. In many cases, particularly toward the end of the war, the manufacturing engineers were involved in all phases of the development process[18]. Operations research techniques such as critical path methods were also developed during this period.

With the growing complexity of engineering design and expansion in the scope of design, engineering organizations became more complex. There were debates whether they should be organized by product or by function. For example, during the second world war, the British government sponsored several studies to compare the design management techniques of US. and British aircraft manufacturers. In many American firms an ad hoc management team was responsible for the project from the inception to the production stage. Organizations were often based on product parts, such as fuselage, airfoils, and engines. In comparison, most British firms were organized on a functional basis. A chief designer was responsible for a project, and the detailed design activities were carried out by different departments, such as aerodynamics, structures, and control systems. In either case, the task of design became further differentiated, with different departments and flows of information back and forth between them. Whereas earlier, a single person could make a change and was responsible for its ramifications, now a single change could effect several departments.

Managing these departments was intrinsically difficult. Unlike production operations, the output of a design team had only one hard milestone, the completion of the design. Intermediate milestones were defined, but operational completion definitions were always vague. Resources and effort required to design different products were estimated by experts who gained their knowledge by working on many similar projects. When new classes of products were designed, only analogies and ad hoc extrapolations were available for estimating design times[19].

---

[17] M.M. Postan, British War Production, 1952, Her Majesty's Stationery Office.

[18] M.M. Postan, D. Hay and J.D. Scott, Design and Development of Weapons, 1964, Her Majesty's Stationery Office

[19] The Rate and Direction of Inventive Activity: Economic and Social Factors, Report of the National Bureau of Economic Research, Princeton

Throughout the last century, in addition to changes in the process employed for designing products, there were also expansions in the technologies available for fabrication. Many new materials were developed. Plastics became available and replaced wood or metal. The cost of aluminum decreased to make it competitive for many uses. New fabrication techniques were developed, including sintering, structural elements made of bent metal, new methods of forging and casting. Each of these changes affected design by expanding the space of feasible designs and requiring changes in how designs were represented. Some of the new fabrication techniques required their own representation of the product. These may simply be drawings with various additions, such as the split lines of molds or they may be special purpose, such as the unfolded form of a bent metal part or the layers of material that must be built up in the use of composites.

In parallel with the advances changing the organization of the manufacturing and design functions within a firm, there were significant changes in how products were marketed. The growth of mass production required new distribution channels. General stores gave way to major department stores, which grew into merchandising empires. The Woolworth, Kresge and Kress five and ten cent stores were the precursors of today's discount stores. The Sears and Roebuck catalog provided urban scale department store access to the many farms and small towns of rural America[20].

By the 1950s, these many diverse distribution channels suggested something recognized earlier in the automobile industry, the importance of market segmentation and product differentiation[21]. Thus emerged the idea of families of products. Advertising became an important means of articulating not only cost differences, but abstract status and image differences between products. As a result, marketing departments began to play a dominant role in the identification of new products.

The Electronic Revolution and the Expanding Role of Design

With the advent of computing in the 1950s and the widened availability of computers from the 1960s onward, manual tasks were gradually automated. Initially, automation took place within the traditional departmental structures, in a one-for-one substitution of manual operations for automated

---

University Press, 1962.

[20] Gordon L. Weil, Sears Roebuck, USA. The Great American Catalog Store and How it Grew, Stein and Day, New York, 1977.

[21] Peter Drucker, The Concept of the Corporation, John Day, New York, 1946.

ones. The strictly analytical aspects of engineering were automated, but the more practical aspects of fabrication design were not. Integrating trade and fabrication practices still required considerable human judgment. Layout and design was still a drawing generation task, but now done with computer-aided drafting. The transition to electronic drafting did not easily address the continued use of previous designs produced manually, on paper. There was no cheap means to convert these for electronic use.

Within management, the development of computation was much more centralized. Data processing departments were organized to manage accounting, personnel, inventory and other activities. Because of the existence of corporate mainframes, it was much easier to integrate these functions. At the same time, engineering computing grew as a departmental activity and thus separately, using mostly mini-computers. Some early proposals for integrated models of a product were made in the 1960s and 1970s, based on central data processing models, but these were limited by the performance of the computers, incompatibility between different data bases, and the restricted ability to interact with users of the information[22].

In the 1970s, computer access moved to terminals connected to a time-shared mainframe and in the 1980s, to local area networks (LANs) connecting personal computers and engineering workstations with mainframes. Networks potentially eliminate the need to have multiple copies of drawings or engineering data. All data need be stored only once, then accessed remotely from any computer on the network. Such single copy storing eliminates the need for physical distribution and reduces the problem of maintaining current updates.

The rapid evolution of personal computers and workstations has allowed them to become more powerful computationally than the largest computer mainframes of the previous decades. The continued growth in computer power has allowed new tools to be developed. Complex phenomena that previously required mockup or model testing eventually became tractable. Important examples are fluid flow simulations, including aerodynamics, other forms of finite element modeling and optimization of complex structures[23]. Design rule checkers and expert systems are taking over certain quality control

---

[22] Vlietstra, J. and R.F. Weilinga (eds.) Computer-Aided Design, North-Holland Press, N.Y. 1973.

[23] For aerodynamic simulation, see Antony Jameson, "Computational Aerodynamics for Aircraft Design" Science, vol 245, 28 July 1989, pp. 361-371; in design optimization, see Gero, J. (ed.) Design Optimization, Academic Press, Orlando, FL. (1985).

and evaluation tasks[24]. Physical models are not completely obsolete, but their use in many industries is disappearing rapidly. CAD systems and many other large application packages can now be operated for only a few dollars per hour, making their accessibility more an issue of user training than capital cost.

At the same time, digital scanning of paper documents and the evolving standards for large format scanned documents, allows even data that was originally on paper to be converted to an electronic format[25]. As a result, all design data is quickly becoming electronic data, with only a relatively few formats. It has become possible to store all information for a project on one or a few magnetic disks. The full implications of this transition are only now being fully recognized.

In the last three decades, the electronic revolution has also had a major impact on manufacturing methods. The development of numerically controlled machine (CNC) tools is considered to be historically the most significant development of the electronic revolution as it affects manufacturing[26]. In the 1980's the capabilities of CNC machines were extended by the development of high level programming languages, allowing the complete set of operations to be designed and tested before the part was produced. Recently, it has become possible to not even address the programming of the machine, but rather have it done automatically, based on a three dimensional model of the part surfaces.

Computer control of machines provided designers with direct electronic access to the production process. For over one and half centuries there had been a separation between the product designer and the producer of the product. The intentions of the designer were presented in a drawing and a machinist acted as a translator of this intention into a physical object. Computers have recovered the link between manufacturing and design. Computer Aided Design (CAD) and Computer Aided Manufacturing (CAM) systems allow engineers to generate alternate designs, carry out a number of analyses such as finite element analysis, generate code for machining the part, and evaluate the impact of the design on the manufacturing system.

The developments in electronics and communication technologies have begun to fundamentally alter the role of manufacturing itself. The new numerically controlled machines combine the versatility of general purpose machines with the efficiency of special purpose machines, thereby shifting the

---

[24] Sriram etal, op.cit, Section on Design Methods.

[25] Eastman, C.M., "New view of CAD: scanning and revision of engineering drawings", J. Information and Image Manag. 19:8 (August,1986), pp. 14-17.

[26] Noble, D.F. , Forces of Production, Alfred A. Knopf, New York, 1984.

focus away from economies of scale to economies of scope[27]. Computer Integrated Manufacturing systems, where the computer "supervises" the operations of a set of numerical controlled machines, provide the flexibility to respond quickly to changes in product mix. Under ideal conditions, these technologies allow manufacturing to play a service role, while design and engineering functions play a central role.

There are several constraints arising from technical and commercial considerations that the design must satisfy. Many critical constraints cannot be evaluated until sufficient progress has been made in the design process. Computer aided design (CAD) and computer aided manufacturing (CAM) tools, rapid prototyping techniques, design for assembly (manufacturing) rules, and high level synthesis tools in electronic design are some examples of the recent developments that enable designers to carry out preliminary analysis of the design, anticipate infeasibilities, prune the alternatives early in the design process, and converge more rapidly to a feasible product. The developments have increased the scope of issues that can be taken into account and the accuracy with which the performance can be predicted during the design development process. Considerations of manufacturability, reliability, maintainability, field service and logistics are just a few of the issues that now arise at the design stage.

In summary, in the last three decades the scope of design has increased enormously. By changing design methods, manufacturing processes, and the nature of information gathered about the market, computers and microprocessors have changed how designs are planned, represented, analyzed and communicated.

Contents of The Book

The papers in this book were presented at a working conference on design management held at the University of California at Los Angeles in 1992. It was organized on the belief that there is a new generation of research issues arising out of the new design technologies being developed. New work in design theory in engineering, new computer based tools, new techniques of distributing and managing engineering information, provide many opportunities to both improve the products and to effectively use resources in developing them. At the same time the growing number of competitors in an increasingly international market place, and an ever increasing number of countries having access to the latest technological advances have contributed to

---

[27] Jaikumar, R., Postindustrial Manufacturing, Harvard Business Review, 64, November-December 1986.

making design an extremely important commercial activity.

The papers deal with a broad range of issues at the interface of engineering design and management. They address new considerations regarding the structure of designed products, based not on performance but rather manufacturing and product family considerations. Work is also presented on the information flow and management in design process and scope, timing and allocation of resources to product design. In more detail, the papers are grouped into five areas:

(1) Decomposition of product development projects. To reduce the complexity of product development, it is fairly common to partition the product into subsystems that can be developed as independently as possible. Product decomposition and management of the interfaces between the subsystems significantly influence the duration and the outcome of the project. Decomposition affects how an organization learns and adapts, and thus has an impact on the long term success of the firm. Another aspect of decomposition is consideration of the use of parts or modules in multiple "families" of products. The development of product families suggests a strong relation between the internal structure of a product, usually the sole concern of engineering, and market issues that are traditionally the province of management. The implications of product decomposition are studied by Lovejoy, Taura, Ulrich and Walsh.

(2) Tools and techniques for preliminary evaluation of designs. Typically there are many alternative approaches for achieving the objective of the design. The alternatives, however, can only be fully evaluated when their design is almost complete. New methods of approximate evaluation are being developed that aid in allowing alternatives to be evaluated early, that also can be shown to have high performance. The papers by Dixon, Karmarkar and Karmarkar, and Prinz review work concerned with tools and techniques for evaluating design for issues that were considered only later in the development cycle.

(3) Interface between design and manufacturing, assembly and distribution. Design decisions have important implications for manufacturing and maintenance. Traditionally, these concerns were reviewed after the engineering aspects of the design were defined, allowing only minor changes in response to these issues. Recently, new computer-based tools have been developed that allow consideration of these issues during initial design. These tools address such issues as the cost of manufacture, the assemblibility of the design, and the efficiency of small lot manufacturing schedules. Ahmadi and

Wurgaft, Barkan and Hinckley, Dixon, and Lee and Billington explore the important interfaces between design and the down-stream activities of manufacturing, assembly, and distribution.

(4) Design of information flows. Design is very information intensive. There is an enormous flow of information between different specialists. Each of these specialists may have their own representation of the product, as input for some form of performance analysis that lies in their domain of expertise. A good information system is not only vital for enhancing the efficiency of the development process, but by documenting the evolution of the design it can also be a powerful learning aid for new members of a design team. Later, such information systems have the potential for supporting the audit of design practices. The articles by Eastman and Shirley, and Prinz provide insights into the management of design information flows.

(5) Determination of scope, timing and duration of projects, and allocation of resources. Determining the set of product development projects to be undertaken and allocating resources to the projects are important managerial tasks. In many industries such as software and consumer electronics, product capabilities are incrementally enhanced in successive generations (or releases) of the product. In these industries defining the scope and the timing of the next generation is a very important decision. The problems of resource allocation, determination of the scope, timing and duration of projects, and management of successive generations of products are analyzed by Adler et. al., Cohen and Ho, Dasu, and Walsh.

Acknowledgments

This book and the conference were made possibly by the generous support of the Center for International Business Education and Research and the Center for Technology Management, both of the Anderson Graduate School of Management at UCLA. We are grateful to Professors. J. de La Torre and E.S. Buffa for funding this project. Jovan Grahovac, Harriet Katz, and Kayle McKee provided invaluable editorial and production assistance.

MANAGEMENT OF DESIGN
Engineering and Management Perspectives

1

EXPANDING THE DESIGN SPACE THROUGH INNOVATIVE DESIGN PROCESSES*

Friedrich B. Prinz
Carnegie Mellon University
Pittsburgh, PA

Acknowledgements: The research for this paper has been supported by the Engineering Design Research Center, a National Science Foundation center at Carnegie Mellon University, Pittsburgh, PA.

Abstract: The subject of design and manufacturing has received increasing attention in practically all industrialized countries. Ever increasing competition has forced most manufacturing organizations to rethink the process by which they create new products and improve existing ones. The following paper attempts to describe the traditional product evolution cycle by reviewing historic as well as modern methods of product design and manufacturing. Several limitations and constraints in the traditional process are identified and related to new methods currently under development at Carnegie Mellon's Engineering Design Research Center, a National Science Foundation Center.*

1. The Product Evolution Cycle

Understanding what needs to be done to position our manufacturing companies in the global market race can best be accomplished by considering the entire product evolution cycle, starting from customer perception or a technical opportunity and continuing all the way down to product disposal and recycling.

The goals are obvious: systems, methodologies, and tools are needed which help produce better products faster. To define the requirements for some

---

* Portions of this paper were presented by the author at the Christopher Hinton Lecture, London, England, 1991 and UCLA's Design Management Conference, September, 1992.

of these let us consider the information flow between various stages of product evolution.

Two key information flows occur during the creation of a product. The first is synthesis (Fig. 1).

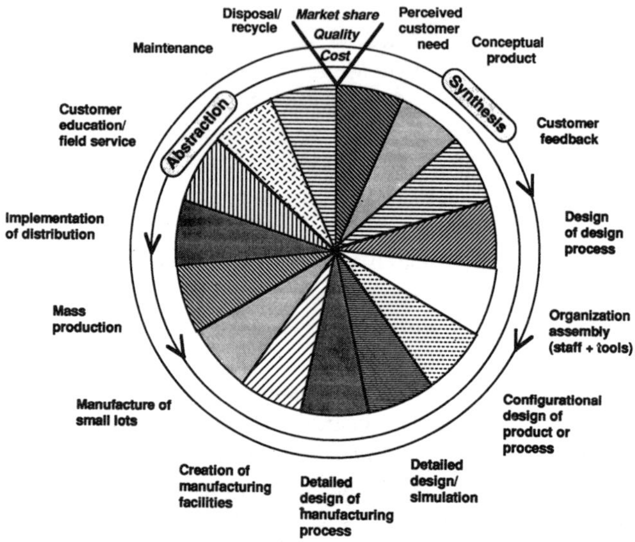

Figure 1

It focuses on the downward information flow in the product cycle. At any stage of the product development there is a large number of alternatives from which to choose. The role of synthesis is the systematic creation of alternatives as the product design and the design process become increasingly more specific. On the other hand, downstream concerns, such as performance and manufacturability, frequently determine constraints on possible alternatives. These constraints need to be accessible at any stage of the product creation cycle where they may influence the process of synthesis. This second information flow represents an abstraction process, since very detailed downstream information usually needs to be presented upstream in a less detailed fashion. A key factor in determining the responsiveness, productivity, and quality of a manufacturing organization is the degree to which these information flows are integrated into the corporate structure.

A concrete example of information flow in an industrial setting: Imagine a car design team synthesizing a new car body. The current manufacturing approach is to spot-weld several hundred pieces of sheet metal together. An attractive alternative is an assembly of lightweight aluminum beams on which plastic panels are mounted. This design is called a "space frame". A design group confronted with this decision needs to understand the essence of aluminum casting and extrusion in order to make an intelligent decision about whether the alternative design is feasible or not. This means simplifying the details of the process to the extent that a nonexpert can take this information as the basis for directing the design one way or the other. The faster and more efficiently the manufacturing process constraints are presented to the designer, the faster and more efficiently they will be able to synthesize a new product.

Decision-support methodologies and tools, which aid in synthesizing and finding constraints early on in the manufacturing process, are key in improving the quality and speed of product creation. Researchers around the world in industry and academia have joined in creating the next generation of tools and methods. It is well known that advances in computer science and communications are pivotal in the development of such decision aids.

The following touches upon only a few of these ideas, tools, and methods. In particular, focus will be placed on that part of the product creation cycle in which a detailed design is reviewed and critiqued by a "decision support system" which has knowledge about certain manufacturing constraints. Consider designing a license plate holder, a relatively simple component for a car (Figs. 2,3).

Figure 2

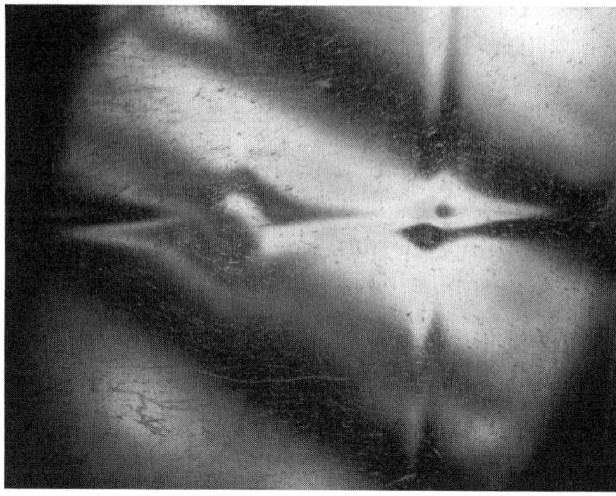

Figure 3

One side of this license plate holder has strengthening ribs to increase the bending strength of the component. On the other side of the same component, you would discover that there is an unevenness in the surface structure, which by design was supposed to be perfectly flat. The reason for this flaw is that the designer who chose the dimensions of the strengthening rib did not apply good design practice and caused the build-up of a volume concentration underneath the strengthening rib (Fig. 4).

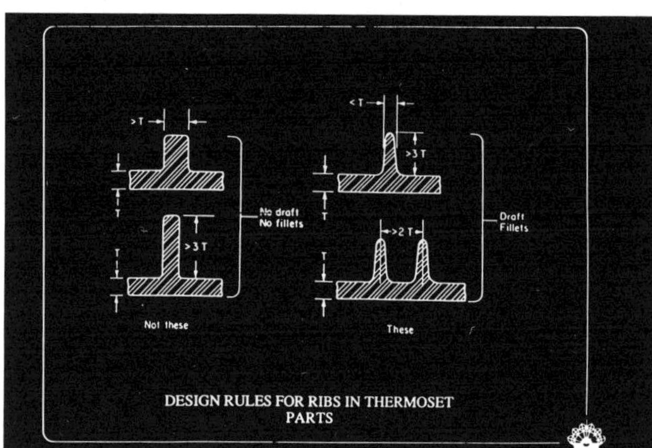

Figure 4

Had some "expert" or some system warned the designer, a mistake which cost tens of thousands of dollars in losses because the mold for the part had to be rebuilt could have been avoided. Even worse, a couple of months in the product creation cycle were lost. An expert system working together with the CAD/CAM system which does precisely that can be viewed in (Fig. 5) [Hall et al., 1990].

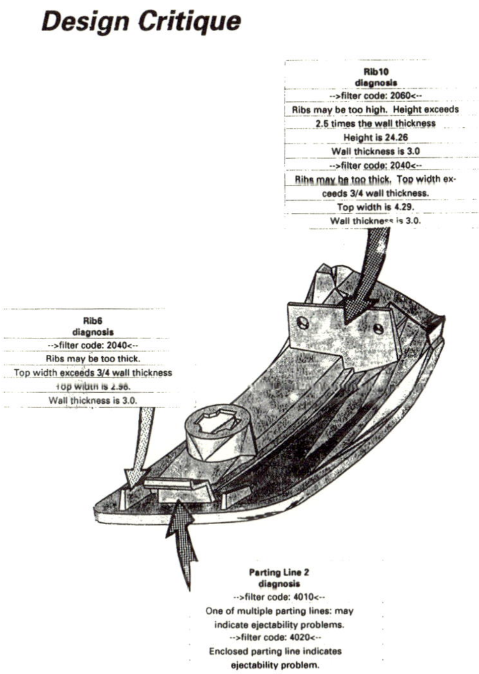

Figure 5

This system recognized certain geometric features which are important from a manufacturing viewpoint. Within that system, the designer receives feedback about whether the dimensions he has chosen are compatible with good manufacturing practice. Decision support tools such as this can significantly reduce errors, cost, and development time. However, the question which still needs to be asked is: Once we have achieved all that, are we going to be better off? Well, certainly we will be at least somewhat better off, because it will take fewer iteration cycles to develop a product, and that will obviously strengthen the market position of companies employing such tools.

From a different perspective, however, if we keep on working over the next few years simply building more and more design critiques, we will consequently have a number of similar systems which prevent designers from making mistakes by informing them of manufacturing constraints. Furthermore, we will be locking ourselves into constraints on manufacturing processes rather than trying to remove those limitations. Preserving existing knowledge may lock us into a certain paradigm which may become increasingly difficult to alter. So, rather than streamlining and automating, let's first look at how we have been making things. Maybe there are techniques which can remove existing process constraints, such as opening up the design space, and allow the creation of products we could have never dreamed of before.

2. Traditional Manufacturing Methods: What They Teach Us

To better understand how things are made today it's always useful to look in to the past. History shows that in manufacturing the components of a product are first formed in one step and then assembled in another. Although the ways of forming a product and the means of assembling it have changed, the basic sequence of forming parts first and then assembling them has always remained. Three historical examples are given to illustrate this sequence and also show how forming and assembly have changed through time. The examples chosen are 1) the creation of an antler spearhead dating back 12,000 to 8,000 years, 2) the making of a common three-legged stool from 15th-century Europe, and 3) Henry Ford's revolutionary use of the assembly line in automobile production. These examples, which illustrate traditional methods of product creation, will then be compared with new ideas and technologies which are currently in a research or pre-competitive stage.

2.1. Prehistoric Antler Spearhead and Stone Tools. The working of bone or antler tools reached a highpoint during the period from 12,000 to 8,000 years before our era. Finely crafted javelins, sewing needles, and harpoons were formed from animal bones, tusks, or antler. Equally impressive are the stone tools used in the formation of these utensils.

The first example, a spearhead made from antler material, was formed with the aid of stone gravers and scrapers. The craftsman first select a slightly curved antler shaft of a male reindeer, killed during the season when the material was most dense. The form of the spear was then cut from the antler with deep incisions made with the graver (Fig. 6) [Daumas, 1969].

This rough "cut" or piece was trued and filed down with a scraper. Following this finishing process, the spearhead was aligned with the aid of "shaft-straightener," which contained a pierced hole and was also made from reindeer antler.

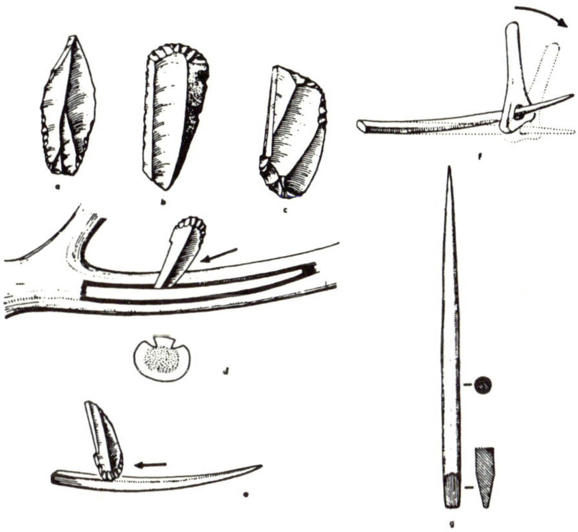

Forming a prehistoric antler spearhead.

Figure 6

One can assume that the completed spearhead was mounted on a thin cylindrical shaft of wood in the assembly stage.

The stone tools used by prehistoric craftsman to fashion the bone or antler implements also represent a high level of technical skill. Early man's knowledge of stone covered a range of materials, such as flint and obsidian on the one hand and coarse-grained granites on the other. The craftsman acquired through tradition a well-developed feeling for these materials, which included knowledge of inherent lines of weakness and faults in stones and how the force of a blow may be distributed. His fine manual and perceptual skills enabled him to strike a stone to achieve a desired shape.

At first, only the simplest hammerstone method of shaping was used. This involved using a water-worn pebble as a hammer to chip off a rough-cut flake from another stone (Fig. 7)

To do this, a tool maker had to know the correct angle of striking the stone to remove a flake at the right point and in the desired direction. Through experience he learned that the angle between his blow and the direction in which he wanted to remove the flake had to be roughly 120 degrees. This "first-cut" flake could then be further worked into a suitable tool.

Using a hammerstone to chip off a rough-cut flake.

Figure 7

By approximately 13,000 B.C., prehistoric man was preparing carefully trimmed stone cores from which long, narrow blade flakes with parallel edges could be struck off quickly (Fig. 8) [Kranzberg and Pursell, 1967].

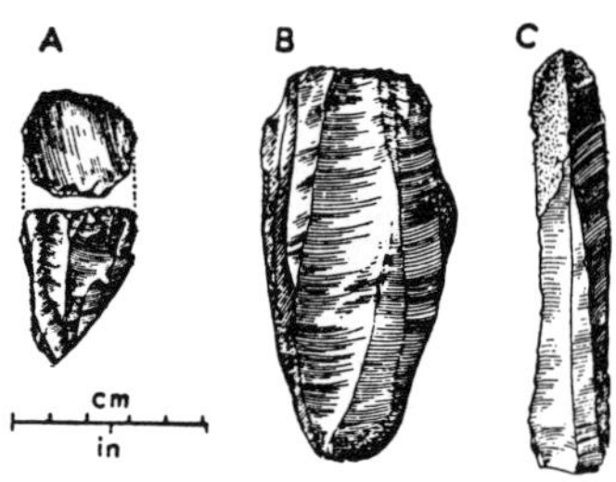

Stone cores (A, B) with semifinished blade (C).

Figure 8

These semi-finished blades could then be made into scrapers or gravers by secondary trimming. By 7,000 B.C., these techniques of working in both bone and stone resulted in composite tools such as this arrowhead in which small flints were mounted with pitch in a shaft of bone.

2.2. Turned Furniture of the Middle Ages. Another example of forming and assembly is the making of turned furniture in the 10th through the 15th centuries, a period of time which marked the highest level of medieval civilization. Two separate crafts were involved in furniture making at this time--the joiners and the turners. The joiners appeared on the scene in the 11th and 12th centuries. Their skill focused on the shaping and fitting together of the component parts of a piece of furniture. The turners, on the other hand, were the oldest craft of furniture makers. They practiced their trade as early as the 8th century B.C. in the Near East, and they made furniture throughout northern Europe from Roman times. Their craft was based on the use of the lathe to cut a furniture part as it rotated or turned on its axis. A simple three-legged stool made around 1430 is an excellent example of forming and assembly, and the advance of the skills and tools at this time. The posts and spindles of the stool were turned on a pole and treadle lathe (Fig. 9).

A joiner and turner at work c 1600.

Figure 9

The furniture part was held between two centers and rotated by means of a cord wrapped around it. The turner controlled the speed of turning through the use of a foot treadle and fixed pole. When the turner stepped down on the treadle with one foot, the piece was rotated rapidly in one direction. When his foot was lifted, the elasticity of the pole pulled back the cord, rotating the part in the opposite direction.

The assembly of the three-legged stool was done with dowelled joints. The seat of the chair was supported by three heavy turned posts, which were held in place with a configuration of spindles. Judging by their appearance in medieval pictures, these chairs must have been produced in great numbers.

2.3. Assembly Line in Automobile Production. The final example of the forming and assembly steps in manufacturing is the contribution made by Henry Ford. The assembly line, which he refined and used on a large scale, embodied the vitality and efficiency of American mass production.

Ford borrowed the concept of continuous movement from the Cincinnati and Chicago meat packing industries who pioneered it in the 1870s. To increase productivity, they hung carcasses on monorail trolleys and transported them past workers in a steady pace with powered movement. Production increased because the pace of work was timed, jobs were specialized, and unnecessary movement by the worker and the product was minimized. Ford caught the idea and introduced it into the production of his Model T with impressive results.

The assembly of flywheel magnetos was the first step. On May 1, 1913 the moving line for their assembly was installed. At first, the magnetos were pushed on rails from one work station to the next (Fig.10) [Singer et al.]. Assembly time dropped from 20 minutes to 13; when powered movement was added, the time dropped to five minutes.

Assembling flywheel magnetos at the Ford Motor Company.

Figure 10

Ford's success with the magneto line led him to apply it to chassis and motor assembly. The average time for the assembly of a chassis, using stationary work stations for each chassis, was 12.5 hours. In a crude trial, Ford attached a chassis to a rope and windlass and pulled it 250 feet past bins of parts. Six assemblers moved with the chassis adding parts along the way. Assembly time was reduced to six hours. Refinements included a chain drive, stationary locations for assemblers, and optimum heights for work stations and parts-supply to reduce worker movement. With these improvements, assembly time was reduced to 93 minutes. Equally impressive times resulted when continuous movement was applied to motor assembly.

It wasn't until the mid-1930s that continuous movement could be extended to the assembly of the automobile body. Before 1935, the body was a wooden frame on which steel sheets were attached (Fig.11) [Hindle and Lubar, 1986]. Building it was slow and costly, and not conducive to assembly line techniques (Fig.12) [Hindle and Lubar, 1986].

Attaching steel sheets to a wooden auto body prior to 1935.

Figure 11

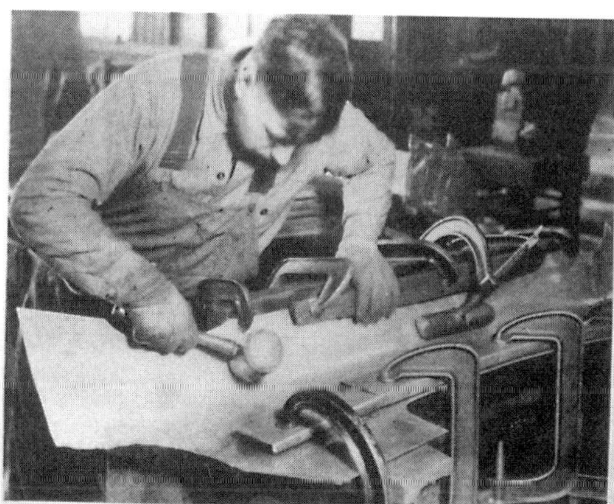

Forming automobile body parts by hand.

Figure 12

The change to an all-metal body allowed manufacturers to form such parts as the hood, fenders, or roof, in presses and then weld them together on the assembly line (Fig.13) [Hindle and Lubar, 1986].

This final link in the automobile assembly line was achieved when the steel industry responded by developing the cold reduction process to produce steel sheets. Steel rolled and annealed in this way could now meet the tremendous stresses of stamping and retain a smooth surface when formed in the presses.

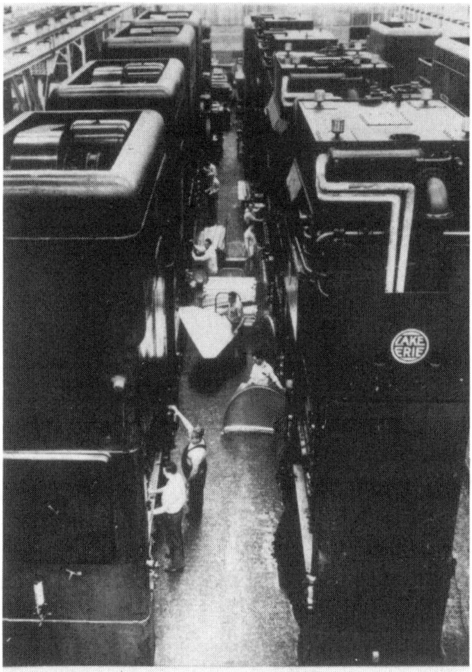

A giant stamping press for forming automobile body parts.

Figure 13

The assembly line, embodying the principle of continuous motion, sped through the automobile industry and other mass production metal-working industries. In effect, it set off a wave of new exploration into ways of achieving productivity gains and set the pace for American manufacturing in the coming decades.

2.4. Historical Conclusions

Some important themes show up in three historical examples of forming and assembling that have been presented.

1) The number of people needed to manufacture a product has increased dramatically. As a result, a division of tasks was instituted. Design and manufacturing, once the domain of one person making a spearhead, were delegated to separate groups in the production of the Model T.

2) The workplace has become larger. In the medieval workshop, the joiner and turner could resolve manufacturing problems on the spot. In the large corporation, communication would prove more difficult.

3) New concepts have been responsible for decreasing production times and controlling output. The carefully prepared stone core resulted in uniform blade flakes that could be quickly produced. Interchangeable parts led to a division of labor and to task-specific machinery that reduced the time to produce a standardized part. Continuous motion of the assembly line speeded up automobile production and permitted its mass production.

These trends have continued into the present, but in each case formation of the components of a product preceded its assembly. Today, we are discovering that this traditional method of forming first and assembling later is making it difficult to manufacture certain products. Imagine the creation of products like "smart structures", components which feel stress, temperature, or which have embedded actuators. While not impossible, the manufacture of such products with conventional manufacturing is difficult where one makes individual components first and subsequently puts them together.

3. New Manufacturing Technologies: A Break with the Past

Let us look at alternate manufacturing technologies which do not impose the well known limitations; technologies where we can shape the components and assemble them simultaneously. And rather than shaping objects through material removal or reshaping, let's consider methodologies in which we incrementally add the material by some deposition process. Technique like thermal spraying are candidate processes. History has shown us that the introduction of new concepts results in very significant advances in manufacturing. Prehistoric man's use of carefully prepared stone cores allowed him to produce uniform semifinished tools. More recently, Henry Ford's application of the assembly line permitted the mass production of the automobile.

3.1. Thermal Spraying. Thermal spraying can be used for any material that melts. One can deposit layers as thin as a few microns, but also as thick as a couple of inches.

The capability of depositing materials in any desirable manner seems very attractive. We may be able to manufacture materials with unique properties, properties which can be designed up-front and which meet certain functional requirements. For example, a component could be manufactured which is hard on one side but tough on the other, or which contains "smarts" in the sense of having the ability to feel and possibly to act. The making of such a component is obviously not a trivial task.

Net-shape fabrication processes can be classified into three categories according to material removal, material reshaping, and material deposition processing. For example, conventional machining processes build shapes by removing material from bulk stock, while casting processes reshape molten material into the desired solidified forms. While deposition processes are relatively new. Currently applied shape deposition processes build parts by incremental material build-up of thin 2-1/2 dimensional layers. A broad class of these processes builds with planar cross-sections, and the growing structure is supported by complementary shaped sacrificial layers (Fig.14).

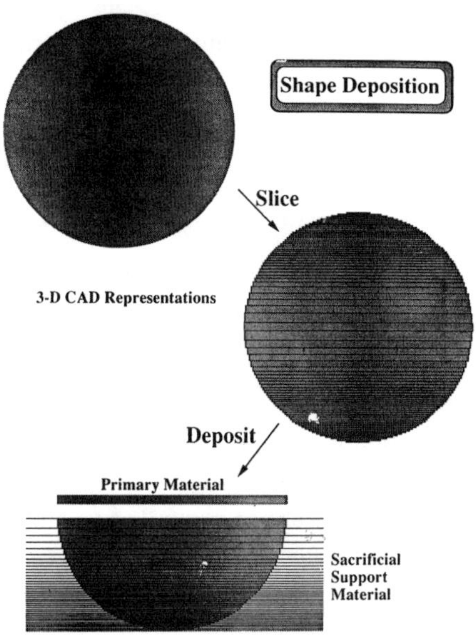

Figure 14

The cross-sectional descriptions are generated by "slicing" three dimensional computer representations into "slices" which may vary in thickness. Several alternative deposition materials and deposition processes are available for building the object, including selective laser sintering, three-dimensional printing, layered object manufacturing, and Cubital's "Solider". The later is similar to 3-D Systems Stereolithography in that both are based upon photopolymerization processes. Stereolithography, however, requires explicit support structure design.

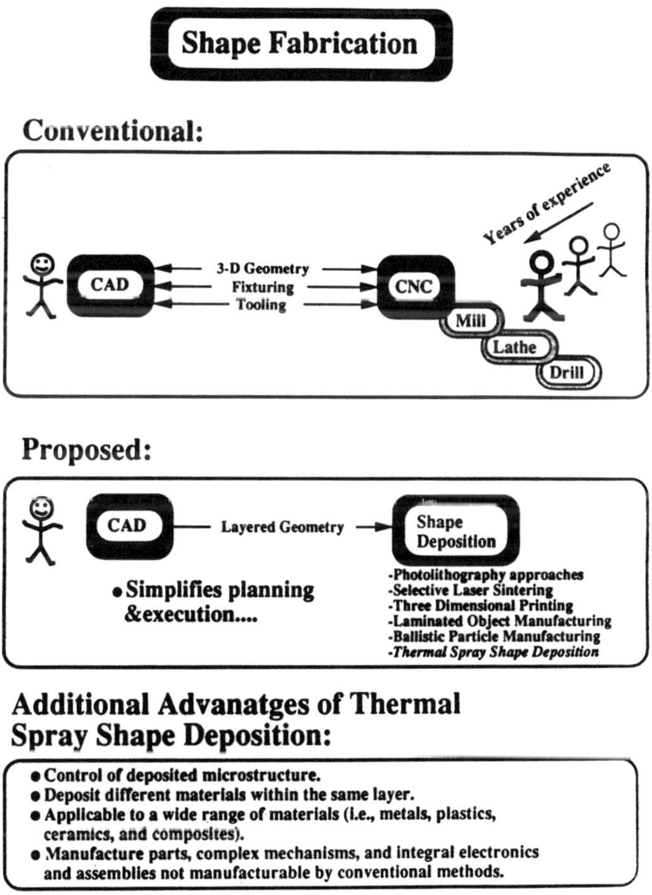

Figure 15

These processes are currently limited to building plastic, ceramic, and wax models. As engineering materials, the plastic and ceramic parts have poor quality and are primarily useful for 3-dimensional visualization and some prototyping applications. However, wax and plastic models can be used as patterns for investment casting and thermal spray processes from which functional shapes are made. Ceramic models can be used directly as shells and cores in casting processes.

A principle advantage of the layered deposition approaches versus conventional CNC machining is the ease and speed with which one can go from design to part fabrication within a CAD/CAM environment (Fig.15).

Planning for CNC machining operations involves several complex problems including reasoning about 3-dimensional geometries, specification of part fixturing, and tooling selection. CNC machining also requires highly skilled, experienced labor. The overall cost, as measured in time and money, for planning executing CNC machining operations is at least proportional to the part complexity and desired precision. In contrast, shaping deposition processes operate on simple 2-1/2 dimensional geometries and do not require either fixturing or tooling information. The part designer can even operate the fabrication machinery. The planning and executive effort for shape deposition is essentially independent of part complexity. Thus, shape deposition facilitates data-driven automation.

Figure 16

A shape deposition process which could directly fabricate functional shapes, composed of several materials, would have a major impact on speeding up product design and on novel manufacturing operations. A shaping deposition process based on thermal spray deposition technologies (i.e., plasma, electric arc, or combustion) would have the potential to achieve the superior material properties required to build functional parts. Thermal spraying is well-established for surface coating and for some near-net shape applications. It can produce dense coatings with desirable mechanical properties associated with the fine grain structure produced by rapid solidification of atomized molten particle beams. Thermal spraying is also suitable for making composite and laminated coatings composed of metals, ceramics, and plastics (Fig. 16).

3.2. Carnegie Mellon's MD* System. At Carnegie Mellon, a system is being developed for rapidly fabricating smart, multi-material products using thermal spraying. The MD* System (for Recursive Mask and Deposit) is a new shaping deposition process in which parts or assemblies are manufactured by successively spraying cross-sectional layers containing several different materials. To create a part, its geometric model is sliced into cross-sectional layers. For each material in a layer, a mask is made that traverses the areas exposed by the mask (Fig. 17).

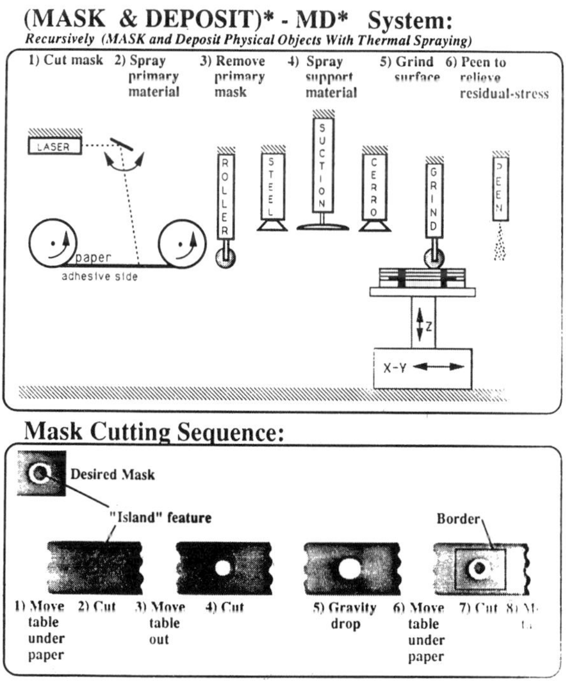

Figure 17

This process is analogous to stenciling for spraying applications. Masks are made from paper stock which is cut with a CO2 laser. Thus, each layer is formed by selective deposition of material, and the complete part is formed by a vertical concatenation of two and a half dimensional layers. A filler material, which can be melted out after completion, is used to separate and support the parts during manufacturing. With the MD* process, the geometry of a part is not constrained by the manufacturing process and the shape and material composition can be changed continuously within a part. High-performance materials can be created, including metals, plastics, ceramics, composites, and laminates. To demonstrate the MD* concept, a prototype zinc turbine blade was fabricated (Fig.18).

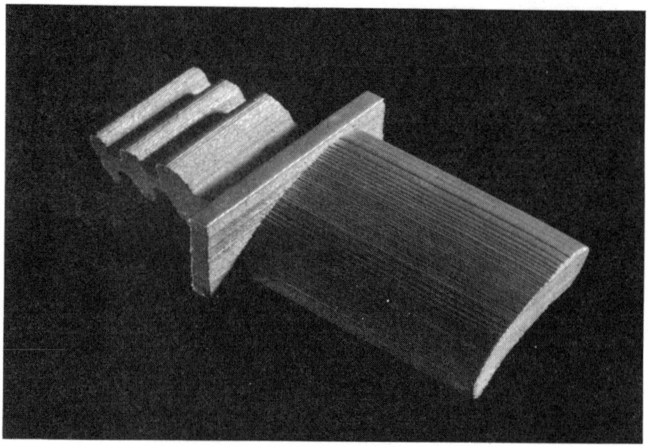

Figure 18

The MD* process would also be capable of creating multi-material structures since masking enables selective material deposition within a layer. For example, this process could create both small and large structures simultaneously such as arrays of miniature strain-gages, thermocouples, and capacitive switches built inside structural elements (Fig.19).

With an MD* process, the fabrication and assembly of complete electronic/mechanical structures, such as computer packages, can be integrated into a single process. Components such as heat pipes, EMI shields, and conductors can be sprayed in place while other discrete components such as chips, external connectors, and batteries can be embedded in the growing structure (Fig.20).

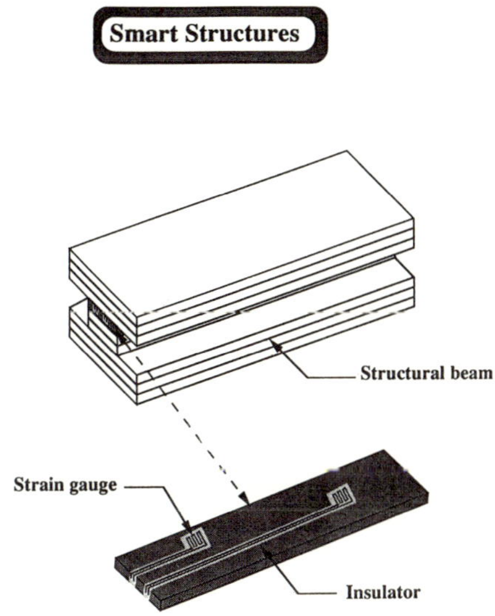

Figure 19

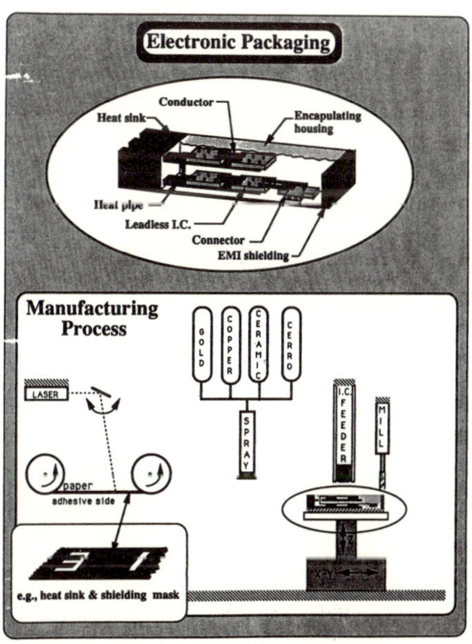

Figure 20

Such integrated electronic/mechanical packages would have several advantages. They should exhibit high reliability due to the elimination of discrete board interconnects, the intimate and absolute mechanical couplings (e.g. the encapsulated connections to ICs should withstand high-G forces), and the elimination of a final assembly process. There should be efficient thermal management since thermal conduits with arbitrarily complex shapes can be fabricated for the specific application and package geometry. Very dense packaging systems could be fabricated since discrete mechanical interconnects and boards are eliminated. Arbitrarily shaped packages can be fabricated to fit nonstandard volumes and to accommodate the stringent space allocation problems associated with, for example, aerospace and automotive structures. And there should be excellent corrosion resistance due to total encapsulation.

To further illustrate the point of simultaneously shaping and assembling, consider the following ball and socket assembly (Fig.21). With conventional manufacturing such a configuration is not feasible.

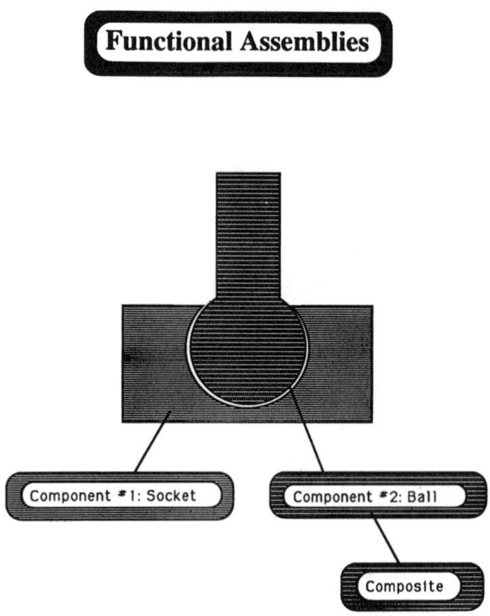

Figure 21

A much more far-reaching example of this technology is indicated by envisioning that a complete product like this car is incrementally shaped and assembled (Fig.22).

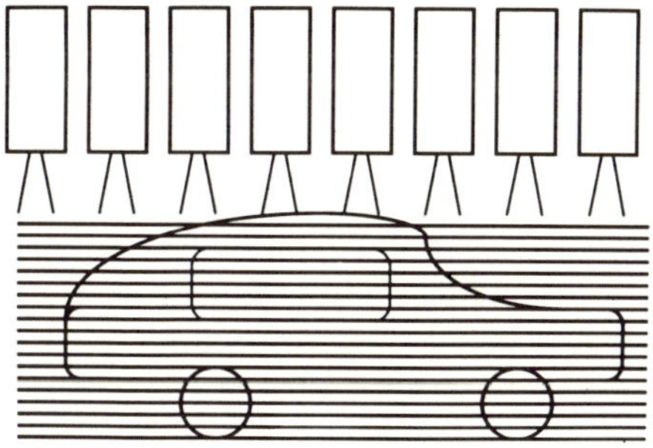

Figure 22

If we imagine that we had MD* spray facilities in many parts of the world, the notion of faxing the car rather than shipping it could be envisioned.

On the more practical side, the MD* technology expands and changes the design space available to the designer of integrated electronic-mechanical parts. The designer will be able to create multi-material, intelligent parts by controlling the material, mechanical, and electrical properties at both a micro and a macro level without the limitations of traditional manufacturing processes such as machining or casting. By giving the designer control over both the shape and material deposition, this process makes possible the creation of designs that previously were not achievable. Many configurations and structures we currently can't make suddenly become possible.

4. Synthesis Computer Technologies Required by Increased Design Space

In order to take advantage of these new possibilities, we need new synthesis methods and tools. New approaches to computer integrated engineering systems capable of reasoning in the unified electronic-mechanical-materials domains are required. Now the creation of possible design alternatives becomes an even more challenging task because of the significantly increased design space.

The following describes some of these emerging computer technologies which either automatically or with user guidance can synthesize structures in two or three dimensions. Conceptual layout is usually one of the first steps in creating a structure. These structures may be in the electrical, mechanical, architectural, or even microstructural domain. Once laid out, another computer tool capable of detailing completes a full-fledged design. For layout, a tool called ABLOOS is described; for adding design detail the use of a tool called GENESIS is discussed.

4.1. ABLOOS: An Evolving Framework and Tool Kit for Layout Design. Layout design deals with many of the complex issues that typically arise in the design of artifacts that have to satisfy specified constraints and are composed of parts that have shape and take up space.

A large (potentially infinite) number of location and orientation combinations are available for placing any single object. In each combination design objects interact in intricate ways through their shapes, sizes, and the spatial or topological relations that exist between them. These characteristics also interact in complex patterns with multiple performance criteria or functional attributes demanded of the artifact being designed. Layout design decisions must simultaneously satisfy global requirements (e.g. usage of space) and local requirements (e.g. adjacencies between pairs of objects); an acceptable spatial arrangement often exhibits a complex pattern of trade-offs.

For these reasons, there is no known direct method which is guaranteed to produce feasible solutions without trial and error for most application domains. Some amount of exploration of the structure, or formulation, of the layout task and search for candidate for solutions is required. However, due to cognitive limitations, human designers do not have the capability of making systematic explorations of alternative arrangements. This shortcoming in human performance has motivated numerous attempts to apply computational methods to layout. What is desired is a structured method for producing multiple alternatives, each of which embodies trade-offs that can be understood, justified, and indicates a range of possible variations within which optimization can take place.

Attempts to arrive at such a method confront the challenges mentioned above; consequently there is a long history of attempts to develop an effective computational method reflecting a variety of representations, system architectures, and planning strategies for layout design. Finding an effective representation to support the efficient generation and evaluation of design alternatives has been a difficult undertaking and has dominated the evolution of the field. The representation must support the creation of a space of possible designs by capturing meaningful differences between design alternatives at a manageable level of detail (or abstraction). Layouts for a given design problem are typically very large; therefore, the representation must allow for the employment of effective planning and search strategies to enable reasonable examination of the best alternatives, e.g., through the evaluation of partial solutions and the incremental specification of designs.

In response to the challenges of the layout task and limitations of other approaches, the LOOS system for layout was developed. It enables the systematic generation of layout alternatives and their evaluation over multiple performance criteria. LOOS utilizes a particularly promising graph-based representation that separates topological issues (spatial relations between objects) from metrical issues (dimensions and dimensional positions of objects) in layout. The representation uses basic spatial relations-above, below, to the right of, to the left--to define the structure or `topology' of a layout as a set of relations between pairs of rectangles. It represents this structure formally through an arc-colored directed graph whose vertices represent the rectangles in a layout and whose arcs represent the spatial relations between the rectangles. The following shows an example in which solid arrows indicate above/below relations; dashed arrows indicate left/right relations, and E represents the minimum enclosing rectangle that is above, right-of, left-of and below all other rectangles in the layout. Using this representation, a set of rules or operations is defined that can generate all possible arrangements of rectangles in a plane by insertion of one rectangle at a time (Figs. 23A-B).

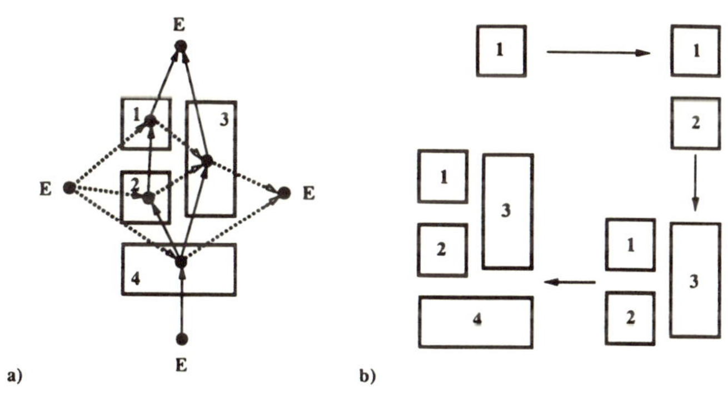

Formal representation and generation of arrangements of rectangles.

Figure 23

The layouts produced by LOOS are loosely-packed arrangements of rectangles; that is, the rectangles are pair-wise non-overlapping, but need not fill a surrounding rectangle. Therefore, the approach is general enough to encompass a broad class of layouts and is useful over a wide range of domains. These rectangular arrangements are given meaning as layouts in a particular domain by attributing the layout objects or components from the domain to respective rectangles. In addition, tests or performance requirements for the layout are attached to these objects, enabling the layouts produced to be comparatively evaluated. Those that are failing requirements may be discarded, while those that show promise can be further developed. The general-and-test design strategy of LOOS has enabled it to produce quality solutions for small but realistic layout problems from a variety of design disciplines [Coyne and Flemming, 1990].

ABLOOS (Abstraction-Based LOOS) builds on the LOOS system for layout, increasing the size of the layout tasks it can handle and enhancing its applicability across domains. ABLOOS allows a layout task to be hierarchically decomposed into subtasks using a representation we call goal-objects (GOBs). Each GOB represents a layout problem at a specific level of abstraction. The layout subtasks are then solved and recomposed to achieve an overall solution. (Fig.24 illustrates the architecture of ABLOOS for the first version of the framework) [Flemming et al., 1992].

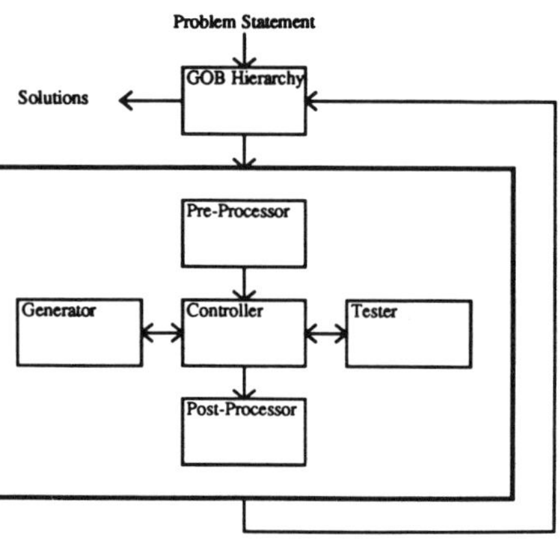

ABLOOS architecture - version 1.

Figure 24

ABLOOS provides a flexible, hierarchical, and interactive tool kit for layout design. The tool kit is comprised of a collection of operations or rules to generate layouts, propagate dimensional constraints, remove and reinsert objects, aggregate and disaggregate objects, and evaluate objects in terms of multiple constraints. The rules work on a formalized, graph-based relational representation. The design process depends on a human computer partnership in which the computer enumerates a space of alternative designs and rapidly evaluates the alternatives. The human-designer creates the evaluation rules and an appropriate decomposition for a given task. The operations in the tool kit may be flexibly combined to yield a variety of powerful methods and strategies for layout design. These range from systematic and exhaustive enumeration of alternatives to incremental interactive design, editing of nearly correct layouts, and redesign.

Applications of ABLOOS have produced high-quality solutions on classes of industrial layout tasks, e.g. the layout of analog computer boards involving 60 components with multiple constraints on their placement. The adaptability of the framework across domains and disciplines has been demonstrated, and included stacked layouts in 2 1/2D space and truly 3-dimensional arrangements. The figures on this side (Fig.25) [Flemming et al., 1992] show sample solutions generated by ABLOOS for the layout of an analog power board.

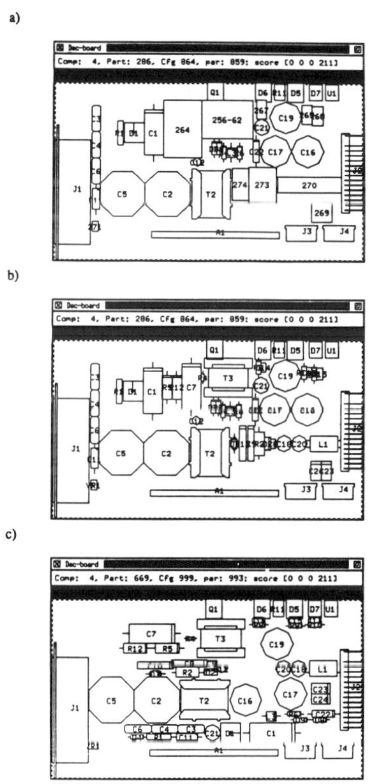

Board placement solutions from top-down power board synthesis.

Figure 25

Figure A shows the final board layout in which all components are placed. Figure B is the same final placement, but with all interior components of aggregated or hierarchical components shown. Alternative board placement solutions, where the preferred component alignment axis is horizontal, can also be generated using the same decomposition strategy; one of these is shown in Figure C.

ABLOOS is currently being extended to handle applications in several domains, including facilities and plant layout computer board and chip layout. Research is underway to incorporate mathematical programming and combinatorial optimization methods into the framework. ABLOOS is also being used as a test bed to investigate methods of knowledge capture, the building of design repositories, and learning within engineering design environments.

4.2. Genesis: A Boundary Solid Grammar. As mentioned before, laying out abstract objects such as boxes does not make a design. All it does is give a spatially feasible configuration of the objects considered. The next step involves incorporating all detailed features of the design, both geometric as well as non-geometric ones. To facilitate this step, it is convenient to introduce

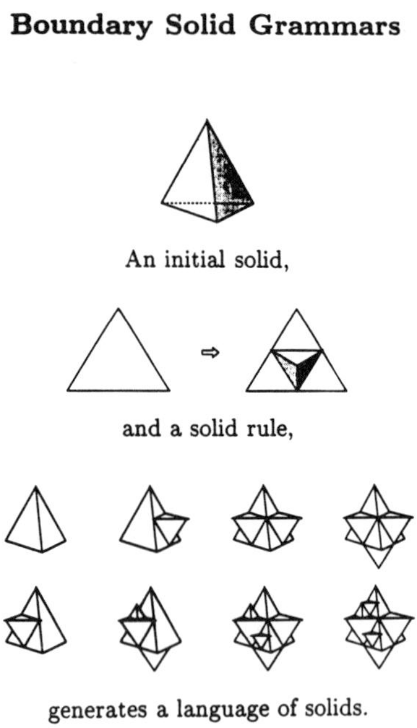

Figure 26

a formal language with a grammar to express the intentions of where to generate what entity in what shape and size, and to determine what other non-geometric entity should be assigned to it (Fig.26).

Solids can be represented through their boundaries. The developed boundary solid grammar developed at Carnegie Mellon/EDRC called Genesis provides a means of generating complex models of rigid solid objects. Solids are represented by their boundary elements, i.e. vertices, edges and faces, with coordinate geometry associated with the vertices. Labels may be associated with any of these elements. Rules match conditions of a solid or collection of solids and may modify them or create additional solids. A boundary solid grammar uses an initial solid and a set of rules to produce a language of solid models.

4.3. Integration of Information Flows and Tools. Integrating and facilitating the top down (synthesis) and bottom up (abstraction) information flow is a key concern in every organization. In an industrial environment there typically are hundreds, and in some selected engineering domains thousands, of such decision support tools in use to create a single artifact, such as an artifact, such as an aircraft engine. These tools obviously encompass a wide variety of modeling, analysis, and simulation tasks, but need to be properly linked to facilitate top down and bottom up information flows. They are subject to permanent change, and one engineer usually works only with a limited set of such tools. What is important to understand is that the integration of such a large number of decision aids is a discipline in itself. Choosing the proper integration frameworks has an important influence on the productivity, and many would maintain also on the creativity, of an engineering organization. Creativity increases with the extent that a productive organization has more time to think.

It is believed that advances in design science in the next ten to fifteen years will increase our understanding of increasingly larger portions of the design process and the design of artifacts. The utility of science-based tools will gradually expand from a few phases of the product creation cycle (detailed design and simulation, detailed design of manufacturing process) to an increasingly larger portion of the total cycle. Most of the tools and methods that have been described here, however, focus only on that detailed design process.

Furthermore, tools based on a scientific understanding will evolve from single designer, single-domain "island of automation" through systems which provide a single designer access to tools in diverse domains (e.g. electrical and thermal aspects of an artifact). This will culminate in integrated networks supporting multiple design teams.

5. Conclusion

This article attempted to traverse through various stages of product development and realization. History shows that the basic manufacturing paradigm for most of our products has not changed. New manufacturing technologies may remove many traditional manufacturing limitations, the consequence of which is a significantly increased design space. The exploration of that new design space will require a new and more powerful synthesis methodologies and tools.

One can be assured that the design research community intends to build an environment in which we can create more freedom for thought, more freedom for the realization of radically different ideas, and more room for individual and collective brainstorming. This environment creates an atmosphere where the unusual and the unconventional is perceived as normal, and where there is not less, but more time for human interaction and negotiation.

REFERENCES

A History of Technology, Charles J. Singer, et al, eds., Oxford University Press, NY, NY.

A History of Technology & Invention, v. 1,2, Maurice Daumas, ed., Crown Publishers, Inc., NY, NY, (1969).

Cooper, Carolyn C., "A Whole Battalion of Stockers", The Journal of the SIA v. 14, Yale University, New Haven, CT (1988).

Coyne R.F. and U. Flemming, "Planning in Design Synthesis -Abstraction-based LOOS," Artificial Intelligence in Engineering. Vol. 1 - Design (Proc. Fifth International Conference, Boston, MA), J. Gero, ed., New York: Springer (1990) pp. 91-111.

Flemming, Ulrich, Can A. Baykan, Robert F. Coyne, Mark S. Fox, "Hierarchical Generate-and-Test vs. Constraint-Directed Search", Artificial Intelligence in Design '92, John S. Gero, ed., Klumer Academic Publishers, Dordrecht, The Netherlands, (1992), 817-861.

Hall, Mark A., Atul M. Sudhalkar, Rajit Gadh, E. Levent Gursoz and Friedrich B. Prinz, "Feature Abstraction in Knowledge-Based Critique of Designs, Issues in Design/Manufacture Integration, A. Sharon (ed.), ASME Winter Annual Meeting, Dallas, TX, DE-V.29 (1990): 43-49.

Heisserman, Jeff and Robert Woodbury, Generating Languages of Solid Models, to appear in the Proceedings of the 2nd ACM/IEEE Symposium of Solid Modeling and Applications, Montreal, May 1993.

Hindle, Brooke and Steven Lubar, Engines of Change: The American Industrial Revolution, 1790-1860, Smithsonian Institution Press. Blue Ridge Summit, PA (1986).

Fussell, Paul S., (private communication), Alcoa Technical Center, Alcoa Center, PA, (1991).

Technology in Western Civilization, V.2, Melvin Kranzberg and Carrol W. Pursell, eds., Oxford University Press, NY, NY, (1967).

Weiss, Lee, F.B. Prinz, D.A. Adams, and D.P. Siewiorek, "Thermal Spray Shape Deposition", Journal of Thermal Spray Technology, V. 1 (3), (1992), 231-237.

2

RECENT DEVELOPMENTS IN ENGINEERING DESIGN RESEARCH

John R. Dixon
Department of Mechanical Engineering
University of Massachusetts
Amherst, MA 01003

1. Introduction: A Brief Modern History of Research in Engineering Design

Engineering design is that phase of the product realization process which, given the output of what is commonly called 'industrial design', adds the information needed for manufacturing a product. That is a *lot* of information. The industrial designers essentially provide only a marketing concept for a product, including some usually incomplete and vague requirements of the market. Manufacturing, of course, requires complete information about the material, dimensions, and tolerances for every part, as well as how the parts are to be assembled into a product. Thus, the amount of information about a product that is added in the engineering design phase is huge -- and the product's cost and quality are extremely sensitive to the decisions made during engineering design.

This being the case, it is remarkable (1) that managers know so little about the processes of engineering design and the knowledge needed to perform them well, and (2) that so little research has been done to understand and improve the processes, and to discover, organize, and generalize the knowledge. Indeed, the typical manager's view of engineering design in product realization is a bit like how people from the East and West coasts of the United States view the nation as a whole. See Figure 1. In coastline minds, that really vast expanse of 'middle America' is shrunk to a sliver. Analogously, many managers -- and this is reflected in such magazines as *Business Week* -- imagine the role of engineering design to be a routine activity between the much more visible (to them) processes of industrial design and manufacturing. But the reality is that thousands of non-routine decisions are made in engineering design that determine product costs, ease of manufacture, and quality.

Similarly, the processes of engineering design have generally been neglected by those who fund research, and hence by those who do research. The very notion of research into 'design theory and methodology' has been viewed with open disdain by engineering scientists, most engineers, and many engineering designers. 'Design theory', in particular, has been considered an oxymoron. Conventional wisdom -- not always wrong, but wrong in this case -- has been that design, including engineering design, is an art and/or skill that can be learned only from the experience of doing it. Denial of the obvious fact that engineering design is a knowledge-based intellectual activity which, therefore, can be profitably researched, remains strongly entrenched.

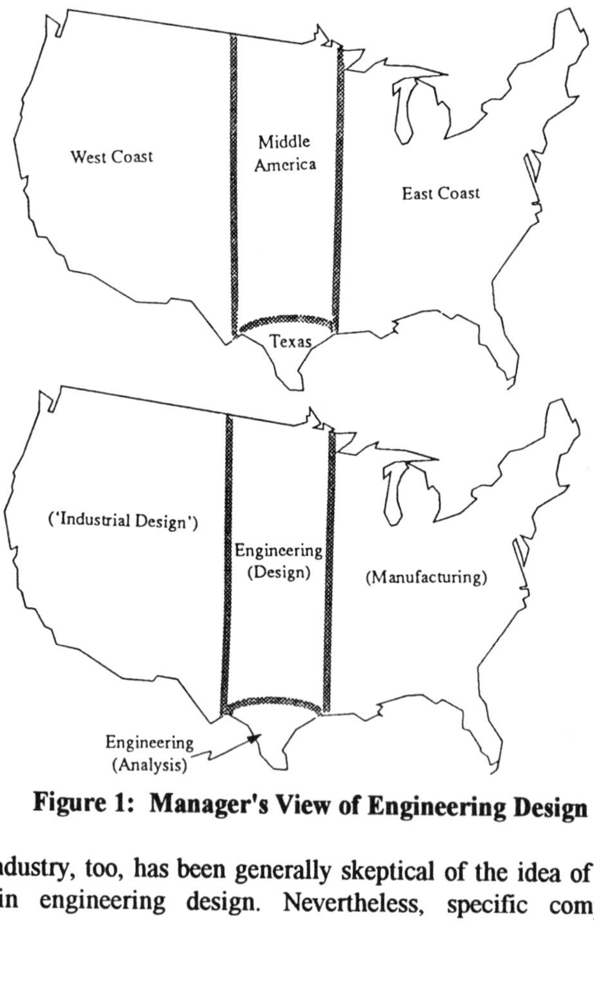

Figure 1: Manager's View of Engineering Design

Industry, too, has been generally skeptical of the idea of fundamental research in engineering design. Nevertheless, specific companies have supported a number of specific researchers on a fairly wide range of specific design-related topics. Such research over the years has produced significant results in such design sub-fields as optimization, computer graphics, design for assembly, finite element programs, and others. Even with these successes, however, the idea of research into more general design methodologies -- and certainly into design theory -- has been difficult to sell to industry.

To his great credit, Dr. Nam P. Suh of M.I.T., while he was Engineering Directorate at the National Science Foundation in the early 1980's, bucked the conventional wisdom and had the vision to create (in 1985) a Design Theory and Methodology Program within a new Division of Design and Manufacturing. With both design and manufacturing declared to be 'emerging sciences', the Division and the Design Program within it were given start-up funding, a division director and program directors, and encouraged to advance the basic understanding of their emerging fields.

The first Design Program Director was Dr. Susan Finger, now Assistant Professor in Civil Engineering at Carnegie Mellon University. Under her direction, a spectrum of research in design was initiated, and a design research community began to form. The program's budget grew to about $3 million. In 1988, I became program director for one year. The budget subsequently peaked at about $4 million, and a very talented and enthusiastic design research community continued to grow. Their meetings and conferences attracted up to 200 attendees from both academia and industry. New journals for publishing design research papers were started, and established journals became more receptive to papers on design.

Of course, the nay-sayers kept the pressure on, and with the departure from NSF of Nam Suh and the few other supporters of design, the effort has fallen on harder times. There has been no solely dedicated program director for some time, and the identity and budget of the program have been substantially diminished. However, even if not well, engineering design is still alive at NSF. And there is new interest at NIST and in a number of industrial firms who have finally recognized the critical role of engineering design in their competitiveness.

In its short and bumpy life, the NSF initiated design research community has accomplished a great deal. There is a growing body of valuable design research literature, and a much deeper understanding of the nature of the knowledge and processes needed to perform design. There is a group of young, extremely bright researchers now at top institutions who have been educated in design and who are working in design. The community continues to grow in enthusiasm and quality if not in size.

Engineering design *is* an emerging science. We are still struggling with development of a basic language and taxonomy of the field, still searching for principles that will give us more generality without sacrificing practicality. We are weak in relating what we know about design processes to the business context in which they take place. We lack analytical and support tools for the important early stages of engineering design. But we are learning, and there is considerable progress to show for our recent efforts. I will try in this paper to outline the boundaries of our understanding in a general way.

2. One View of the Product Realization Process

To provide a context for this review, it is helpful to relate engineering design research to the stages of the product realization process. Figure 2 shows the engineering designer's view. Note the location of engineering design between industrial design and manufacturing. Note, too, the supportive, staff-like role of what most managers call 'engineering' -- that is, engineering *analysis*.

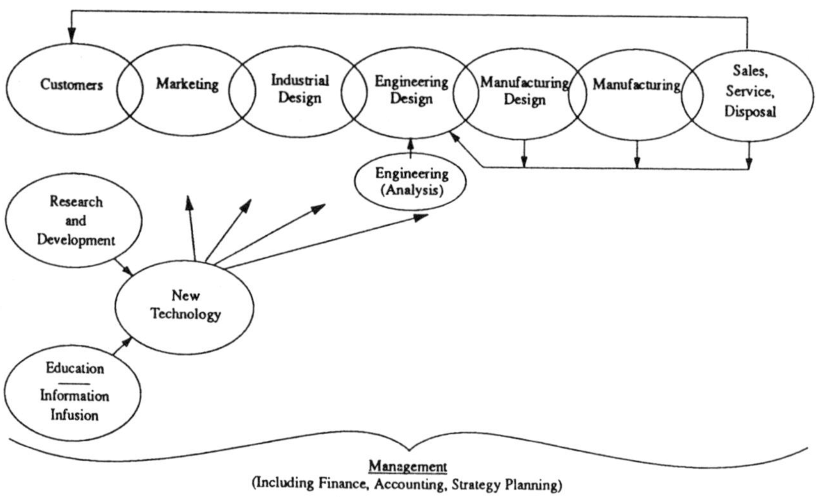

Figure 2: A View of the Product Realization Process

Another view of this total process, with the major stages of engineering design added, is shown in Figure 3. The stages of engineering design shown -- conceptual, configuration, parametric, and detailed -- are not disjoint in practice or theory, but the distinctions are very useful in

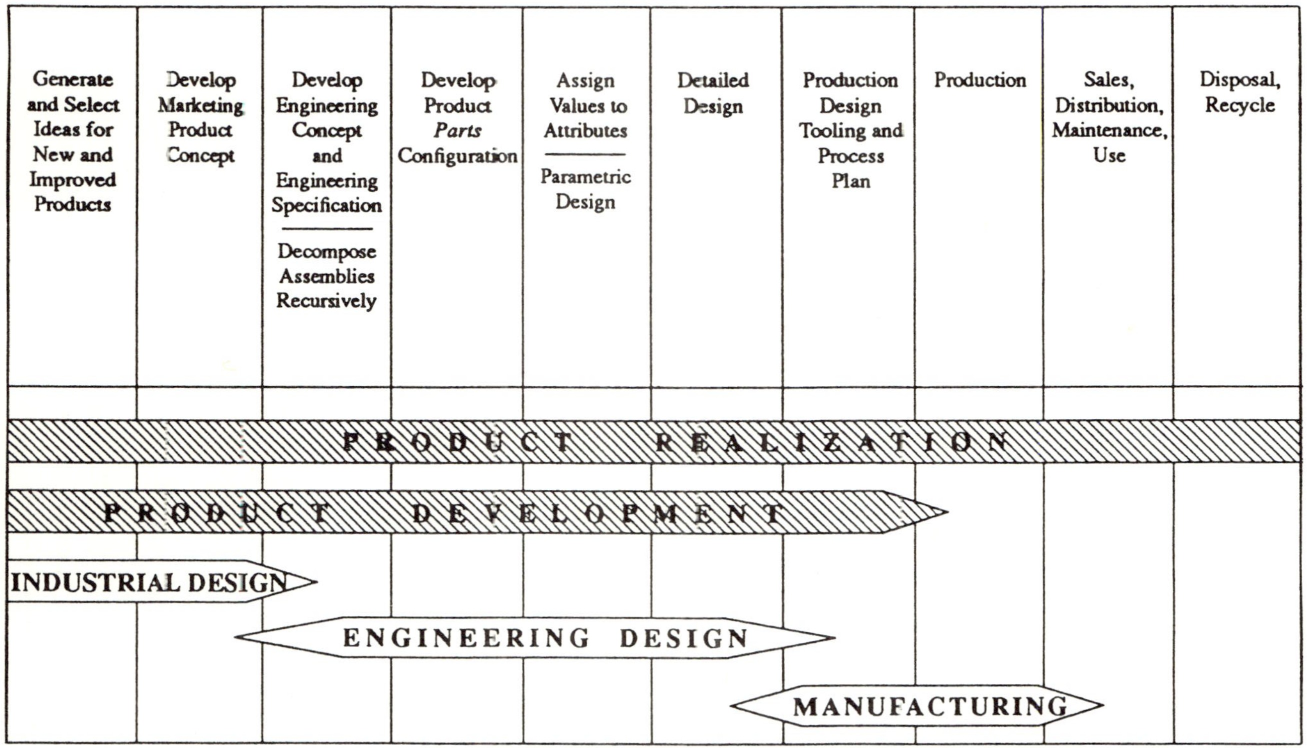

Figure 3: Major Stages of Product Realization (Not to Time Scale)

describing research activities. Moreover, design methodologies -- at least those available today -- tend to be applicable primarily to one or the other of these stages.

3. Three Research Taxonomies

In addition to the context and structure provided by referring research activities to the product realization process, it is also helpful to be able to refer to one or the other of two basic taxonomies of design research. One, from (Finger 1988), organizes the research into six broad areas:

1. Descriptive Models of Engineering Design
2. Prescriptive Models of Engineering Design
3. Computer-Based Models of Engineering Design
4. Languages and Representations in Engineering Design
5. Analysis to Support Engineering Design
6. Engineering Design for X (where X stands for assembly, manufacturing, reliability, disposal, and all the other issues engineering designers must consider as they make design decisions)

Prototypical examples of each of these categories of research are:

• *Descriptive models*: Protocol studies of the cognitive processes employed by designers as they design;

• *Prescriptive models*: Axiomatic theories of design (Suh 1988);

• *Computer-Based models*: Computer models that enable experiments with an iterative redesign model of design; The 'Dominic' series developed at the University of Massachusetts is an example of such an algorithm.

• *Languages and Representations*: Features have been introduced as a more abstract and useful representation for the geometric aspects of parts and as a basis for recording other information as well;

• *Analysis to Support Design*: More intelligent designer oriented interfaces to analytic models such as finite element programs;

• *Design for X*: Design for Assembly is a prime example.

Another taxonomy of engineering design research to which we can refer is the one developed in a National Research Council (NRC) study entitled *Improving Engineering Design: Designing for Competitive Advantage* (National Research Council 1991). The Report recommends a research agenda organized as follows:

A. Development of Scientific Foundations for Models and Methods
   a. Computer Representations and Information Models
   b. Organizing Design Knowledge
   c. Synthesis: Conceptual, Configuration, and Parametric Stages
   d. Tolerance Synthesis

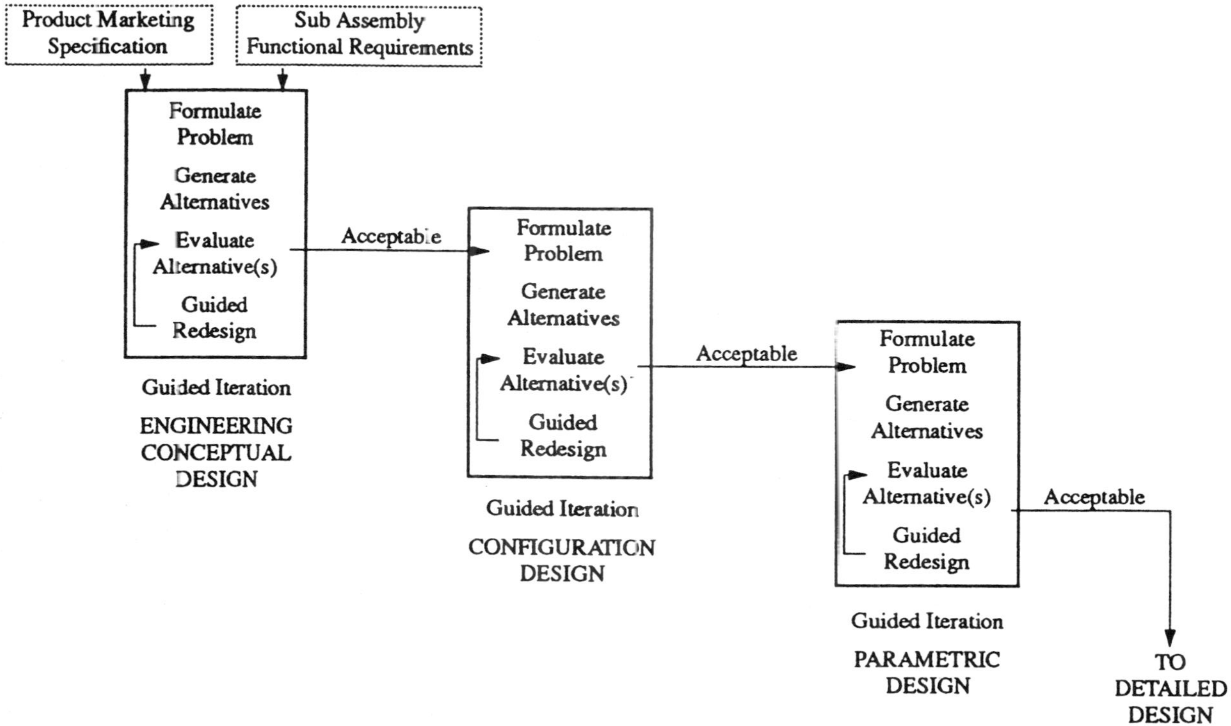

Figure 4: Guided Iteration Is Used at Each Stage of Engineering Design

B. Development of Supporting Tools and Methods
   a. For Computational Prototyping
   b. For Rapid and Meaningful Physical Prototyping
   c. For Designing for X
C. Development of Models Relevant to the Business Context of Design
   a. Early Quality-Cost Models
   b. Organization and Communication Models
   c. Innovation and Commercialization of New Science and Technology

We can also organize the progress, or lack thereof, in terms of a generic design process. Despite objections to the contrary, and notwithstanding the complexities and huge number of factors that mask it, we do know (even if we have not yet admitted it) the basic structure of all engineering design processes (my opinion). The process at an abstract level is essentially the same for each of the phases of engineering design, and their sub-processes. It is fairly accurately described as 'guided iteration'. The process (see Figure 4) is:

- *Formulation* of the problem at hand in terms of the information to be added and the larger technical, economic, and business context in which the problem exists;
- *Generation* of alternatives solutions that add the required information;
- *Evaluation* of a selected set of the alternatives, supported as much as possible by analytical procedures that provide the quantitative information on which evaluation can be based. If no alternative is fully acceptable, then
- *Redesign* *guided* by the results of the evaluations.

This abstract description of the guided iteration process makes more explicit the crucial nature of evaluation than do the taxonomies shown above. Indeed, the ability to evaluate trial alternative designs is the keystone to successful design at every stage and in every design problem, large or small. Analysis *supports* evaluation (by providing quantitative information) and is therefore important to design, but design decisions require further evaluations which are value laden, multifaceted, and goal directed.

With these contexts to refer to, let us see where the field of engineering design research stands.

3. Engineering Conceptual Design

Engineering conceptual design is a hot subject at this time. See, for example, the recent Proceedings of the ASME Design Theory and Methodology Conference (ASME 1992). In the conceptual design phase, designers determine the physical principles by which a product or machine and its parts will perform their various functions. The basic materials and manufacturing processes to be used are selected. Products are decomposed into their major functional sub-assemblies, and those sub-assemblies into sub-assemblies, and so on recursively until the level of individual parts or standard components is reached. See Figures 5 and 6.

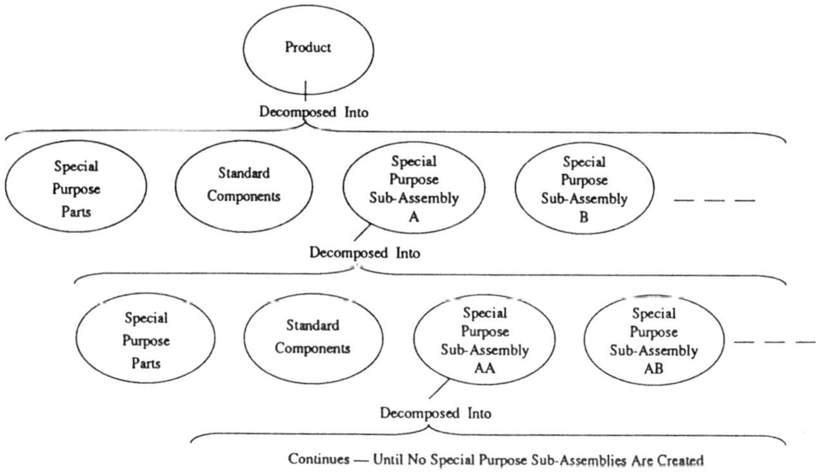

Figure 5: Engineering Conceptual Design Is Done Recursively

We know little of a formal nature about this decomposition process. In products with many parts, we know the process is *extremely* complex. Even in smaller products, it is intellectually complex. The ways in which parts, components, and sub-assemblies are coupled (not just their physical and geometric connections and relationships, but also their energy, force, and information inter-relationships) can not really yet be represented in ways that support the design process. Moreover, there are many design related activities to be represented, each needing different information about the partly designed product. There are no prescriptions for how best to perform the decompositions, except that we know it is best to minimize the complexity of the various inter-relationships. (Of course, the prescription to keep a design simple is not new.)

There are beginning to be some supportive computer tools, at least for certain kinds of products. ICAD Corporation in Cambridge, Massachusetts, markets and supports an 'intelligent' assembly design system that does an excellent job of enabling designers to express their desires for decomposing products, recording and maintaining certain relationships among the sub-assemblies and parts, and bringing catalog information into the design.

A method -- called Pugh's method -- for evaluating conceptual designs has been developed by Professor Stuart Pugh in England [Pugh 1990]. The method essentially focuses the attention and records the subjective evaluations of an interdisciplinary team of designers and others on the criteria which determine the quality and efficacy of alternative concepts.

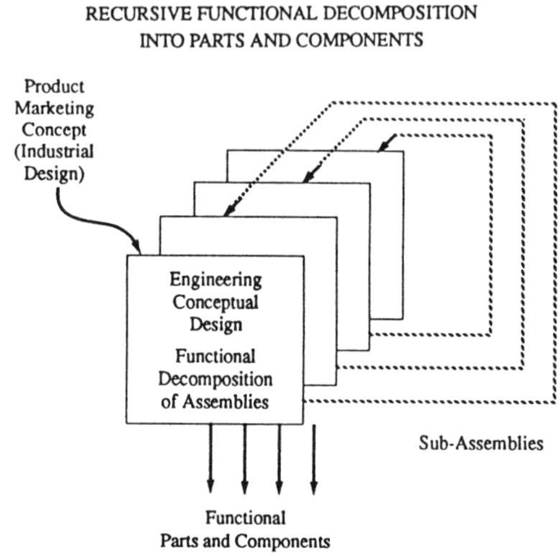

Figure 6: Another View of Conceptual Engineering Design

Bond graphs have been experimented with as a way to represent the functional couplings of a decomposed assembly, and even as the basis for a method of generating decompositions. The formalism appears to have promise, but more work is needed.

4. Part Design

Most parts are thin-walled, and thin-walled parts are composed of features: that is, of walls, intersections among walls, and 'add-ons' to walls like bosses, holes, ribs, etc. The selection of the features in a part and the connectivity among the features is based on simultaneous consideration of qualitative physical reasoning and manufacturabilty concerns. Qualitative physical reasoning, for example, leads us to the I-like arrangement of walls in an I-beam. Manufacturability concerns lead us, in an injection molded part for example, to try to avoid holes or bosses in a direction perpendicular to mold closure since these would require more expensive side action molds.

We use qualitative physical reasoning in only informal, essentially unconscious ways to generate and evaluate the features and their arrangement in parts. For the most widely used processes (injection molding, die casting, and stamping) we have considerable information about part design for manufacturability, and can actually estimate tooling costs from only the information about which features are present and how they are arranged. That is, we do not need to know exact dimensions to predict tooling costs, and hence we can make this evaluation early in the design before parametric design has been performed. For processing costs, we generally need the actual dimensions.

To make part design more efficient, we need feature-based design systems. Much of the basic research work needed to develop these systems has been done, but practical systems are not yet available. They will not be available for some time, either, because there is a huge software engineering task to be done to complete their development. And there is no one to do it. I call this the 'software gap.' University research can generate new ideas like feature based design systems, and carry the research to the point of proof-of-concept operating systems. Beyond that, however, the work to be done does not generally qualify for academic thesis credit, and so cannot be done in the universities. In the U.S., manufacturing firms do not do this kind of software development either, and the software firms simply do not have the capital to do it. This gap between proof-of-concept and finished software is big and expensive to bridge. It takes either a consortium of companies or else a government agency like NIST to support it with the funds and time needed. Mostly, very good university research gets someone a PhD, is published, and lies fallow in the software gap for years, decades, or forever.

5. Parametric Design

When it comes to assigning actual numbers to things -- that is, parametric design -- we are in relatively good shape. Optimization methods and computer programs are very powerful. Taguchi methods are available when robustness is an important design issue. We can do basic shape optimization for structural efficiency of parts pretty much automatically. And we have a number of supporting analysis tools like finite element programs, simulation programs, and a few rapid physical prototyping technologies.

The situation is not as good when it comes to tolerance design. Tolerances are a very sensitive interface between function and manufacture. While we can analyze tolerances fairly well, synthesis methods are less well developed. Taguchi's methods provide one way, and a promising method using fractals has recently been proposed (Srinivasan 1992). This is the kind of exciting new idea that comes from basic research in engineering design.

6. Design for Assembly

The most fully developed design for X topic is design for assembly. The work pioneered by Geoffrey Boothroyd has given us easy to use, effective methods for guiding and evaluating part design for ease of assembly.

We learned from this research that it is generally better to design parts and assemblies correctly than it is to try to find ways to assemble parts that have been poorly designed. Another generic result is that reducing the number of parts is crucial to reducing assembly (as well as inventory) costs. How great can the complexity be? It appears that a great deal of complexity can be justified. That is, whenever possible, it appears to be better to use a single complex part than to assemble two more simpler parts. Karl Ulrich (Ulrich 1990) has shown, however, that if the complex part creates critical delays in the design cycle, this general conclusion may not be true.

7. Other Current Research Topics

Quality-Function Deployment. A method called quality-function deployment [Hauser and Clausing 1988] has been developed for relating the characteristics of products that purchasers desire to the more measurable and observable engineering characteristics that designers can control. A technique called the House of Quality [Hauser and Clausing 1988] is available for this purpose. The method is also helpful in performing competitive benchmarking and in improving existing products. The method continues to be improved.

Mechatronics. The integration into so-called 'smart' products, of electronic, computer, optical, and mechanical elements, is of considerable interest. This is, at one level, simply a special case of 'systems design' or 'conceptual design'. However, it is very special and of sufficient importance that it is receiving some research attention, especially in the United Kingdom.

Information Models. To serve a more concurrent design process, it is desirable to have a single information model of a product that all parties can have access to. This is another 'hot' subject at this time. What information should be included? How should the data be represented? How should modification or additions to the information be controlled? These and other questions are being addressed, but are far from solved.

Design for Recycling and Disposal. One of the newest X's in design for X is disposal. How can products be designed so that they can be dis-assembled economically and the parts recycled or disposed of in an environmentally safe manner? A little research has begun to appear on this subject.

Accounting for Design. In most firms, design is viewed by both financial and management accounting systems as an expense. It is, however, far more realistic to view design as an investment. To my knowledge, there is little research being done to determine how this could be done. However, so long as managers view design as an expense rather than as an investment, there is little hope that the true impact of design on competitiveness will be recognized.

8. Conclusion - From Art to Science

Engineering design overlaps and connects industrial design with manufacturing. It adds a great deal of information about the product, and involves a great many decisions that impact the cost, manufacturabilty, and quality of products. The field of engineering design has traditionally been treated as a collection of special devices or methods, and these have been the subject of applied research. Only in the past decade has research funding and attention been directed to more fundamental (i.e., general) methodologies and principles. This more basic research was initiated by NSF, and though the NSF interest has unfortunately waned, the work continues with a cadre of young researchers spawned by the program. Industry interest is growing, too.

The current effort, however, lacks focus. Though very good (even brilliant) work is being done in various places, and published in a variety of journals, there is not nearly the needed scope of research needed for design knowledge and capability to make the contribution it could and should to U.S. manufacturing competitiveness.

Engineering design is, as Nam Suh said a decade ago, an emerging science. Strange it is that the embryonic science of design should be viewed by some with disdain. Design is clearly a sophisticated intellectual process of critical importance to civilization, as important as any of the professions to the survival and well being of the society. That we yet know so little about it is sad enough; to argue that we cannot learn more through research is dangerously Luddite.

In any case, the research continues and will continue. It is gradually adding to our understanding and bearing practical fruit as well. As design progresses from art to science, the engineering design process in practice will become more efficient, more effective, and ever more researchable.

REFERENCES

ASME, Proceedings of the Fourth International Conference on Design Theory and Methodology, Scottsdale, AZ. September 1992.

Finger, S. and Dixon, J. R., A Review of Research in Mechanical Engineering Design, Part I and Part II, Research in Engineering Design, Volume 1, Numbers 1 and 2, 1989.

Hauser, J. R. and Clausing, D., "The House of Quality", Harvard Business Review, May-June 1988, pp 63-73.

National Research Council, Improving Engineering Design: Designing for Competitive Advantage, National Academy Press, Washington DC, 1991.

Pugh, S., Total Design, Addison-Wesley, Reading, MA.,1990.

Suh, N. P., The Principles of Design, Oxford University Press, Oxford, U.K., 1988.

Srinivasan, R. S., and Wood, K. L., Fractal-Based Geometric Tolerancing for Mechanical Design, Proceedings of the Fourth International ASME Conference on Design Theory and Methodology, Scottsdale, AZ. September 1992.

Ulrich, K., Proceedings MI '90, ASME Conference on Manufacturing, Atlanta, GA, 1990.

3

NEW PRODUCT DESIGN STRATEGY ANALYSIS: A MODELING FRAMEWORK*

Morris A. Cohen
Jehoshua Eliashberg
Teck H. Ho

The Wharton School
University of Pennsylvania
Philadelphia, PA 19104-6366

Abstract: The development of new products inevitably leads to tradeoffs among product performance, time to market, and development costs. In addition, the unit cost of the product, which is determined by the product performance, is also affected by the design process. In this paper, we present a model framework that examines the interaction of these factors. We derive several managerial insights that have important implications for new product strategy planning. The model framework is applied to evaluate five industry practices: 1) the traditional approach (iterative), 2) the target costing approach, 3) the target performance approach, 4) the target timing approach, and 5) an approach based on minimizing break-even time. Each practice is a restricted case of an optimal procedure which considers all factors simultaneously. Our model framework allows us to benchmark each practice against the optimal procedure with respect to the size of the development team, time to market, new product performance level target, and unit cost of the product.

---

* This research is partially supported by a grant from the Center for Manufacturing and Logistics Research, The Wharton School, University of Pennsylvania and NSF Grant #522571.

1. Introduction

Competitive pressures and the hastened pace of technology development have increased the strategic importance of new product development. Most companies have increased their product diversity and are placing greater emphasis on being responsive to diverse customer needs. Success in the market now depends on a firm's ability to deliver a high performance product, at a competitive price, in a timely fashion. These success factors (i.e. time to market, product performance, unit cost of the product, and development costs) may conflict, however, and firms must consider the tradeoffs among them. Indeed, many practitioners suggest that a framework that examines these tradeoffs explicitly is central to the strategic management of the new product development process (Smith and Reinertsen, 1991). In this paper, we introduce such a framework and discuss its implications for new product strategy planning. This modeling effort is part of a long-term project to develop models and procedures to facilitate strategic management of new product development.

Of the four success factors, product performance has no common definition among researchers. For example, the operations management (OM) researchers posit that "quality is free", implying that the unit cost of a product can be reduced by increasing the product's quality which is related to the conformance of the product to its expected design performance (Crosby, 1979; Juran and Gryna, 1980; Porteus, 1986). On the other hand, marketing scientists hypothesize that "product performance is costly" and that the unit cost of a product increases with its level of performance (Mussa and Rosen, 1978; Moorthy and Png, 1992). In this paper, we adopt a composite view of product performance by treating it as a random variable with mean (μ) and variance (V) (Karmarkar and Pitbladdo, 1992). This view of product performance integrates the two streams of literature. The OM's view of product performance is captured by the variance term with a low variance implying high product conformance. The marketing's view of product performance is gauged by the mean with a high mean indicating a high performance class. Therefore, the overall relationship between unit product cost and product performance is such that the cost increases with both the mean and variance of performance In addition, our definition of product performance is based on customer perception in the marketplace and thus can be influenced by advertising, brand name, and customer service practices even after the product is launched. Conformance can also be improved during this market stage by investment in continuous improvement of the manufacturing process.

Besides adopting a broader view of product performance, our framework is distinct in several aspects. First, it is the first formal framework that examines the six tradeoffs among the four success factors simultaneously. Previous work (e.g. Mansfield, 1988; Graves, 1989) examine only the tradeoff between time to market and development cost. Second, we model the new product development process as a *multi-stage* performance improvement process. This allows us to study how development and time resources should be allocated across development stages. Third, we adopt a *life-cycle* perspective to new product development process. Both the cumulative costs and revenues of the new product over its entire life cycle are considered. The framework develops in-depth insights on one of the three dimensions of a general framework for making an organization innovative proposed by Cohen and Dougherty (1992).

The paper is organized as follows. In section 2, we describe the framework. We use optimal control theory to characterize new product development policy in terms of time and resource allocations between development stages. We state several managerial insights derived from the framework as a set of testable hypotheses in section 3. In section 4, we show how the framework might be used to analyze the five design approaches which are used by companies in developing new products. The managerial implications of each design approach are discussed.

2. Model Framework

New product development is conceptualized here as a multi-stage product performance improvement process. Figure 1 shows the general structure of the modeling framework. The framework consists of three sets of variables and two sets of functional relationships.

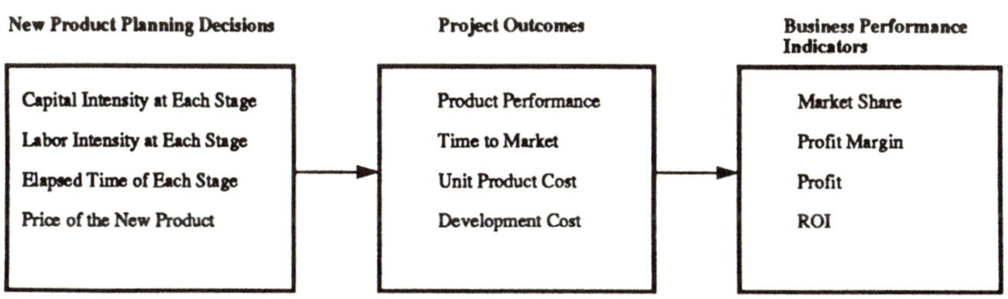

Figure 1: General Structure of the Modeling Framework

For a particular new product development project, one needs to consider:

- New Product Planning Decisions

Intensity level of capital utilized at each developmental stage, K_i.

Intensity level of labor (engineering or sales) utilized at each developmental stage, $L_i(t)$, over time

Elapsed time for each developmental stage, τ_i

Price of the product at time t, p(t)[1]

- Project Outcomes

Product performance in the market at time t, denoted by the random variable Q(t) with mean and variance, $\mu(t)$ and V(t)

Time to market, T_p

Unit product cost at time t, c(t)

Cumulative development cost at time t, TC(t)

- Business Performance Indicators

Market share at time t, S(t)

Profit margin of the product at time t, r(t)

Cumulative profit at time t, $T_\pi(t)$

Return on investment at time t, ROI(t)

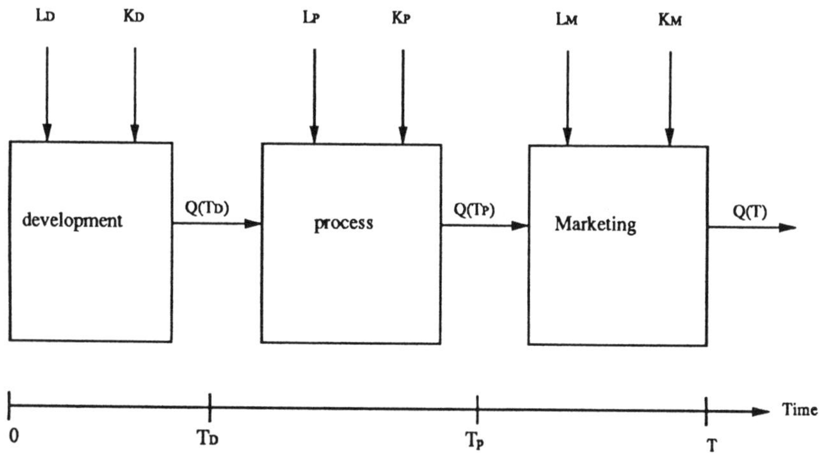

Figure 2: A Three-stage New Product Development Process

Figure 2 illustrates how resource inputs are transformed into product performance via a three-stage product performance improvement process. The Development stage includes concept generation, product design, and

---

[1] In many cases firms are pure price takers in competitive markets and hence we assume that price is exogenously determined.

engineering analysis. The Process stage consists of process analysis and design, and prototype production and testing. The Marketing stage encompasses activities associated with volume production, advertising, and selling of the product. Labor and capital inputs, $L_i(t)$ and K_i , to each stage represent various resources required to carry out the development activities. A more detailed version of this model could consider more detailed classes of inputs (e.g. engineering, designer, production, and marketing labor). Since the window of opportunity is finite, a day spent in the Development or Process stage means a delay of one day in the Marketing stage.

We make some assumptions about two sets of functional relationships. First we assume that the rate of performance improvement at each stage i is a *Cobb-Douglas* production function (see Bohem, 1982; Griliches, 1984). That is, the rates at which the mean and variance of product performance are enhanced are proportional to $L_i^{\alpha_i}K_i^{\beta_i}$ where α_i and β_i are labor and capital productivity parameters. We assume diminishing returns to both the labor and capital investments so that α_i and β_i are both less than 1. Second, we assume that the multi-stage development process links the performance improvement functions in an *additive* manner. Improving product performance is analogous to climbing up a performance ladder (Grossman and Helpman, 1991). The firm is also assumed to have an infinite number of improvement opportunities so that the increase in the product performance can be represented as a continuum. Based on the two assumptions, mean and variance of the product performance at time t in $[T_{i-1}, T_i)$ are given by:

$$\mu(t) = \mu(T_{i-1}) + \int_{T_{i-1}}^{t} \delta_i \; L_i^{\alpha_i} \; K_i^{\beta_i} \, ,$$

$$V(t) = V(T_{i-1}) - \int_{T_{i-1}}^{t} \eta_i \; L_i^{\alpha_i} \; K_i^{\beta_i} \, ,$$

where δ_i and η_i are proportionality constants[2]. Third, we assume that there is industry price leadership or the industry is regulated such that product performance is the main competitive tool. We use the attraction model to determine the market share of the new product (see Bell, Keeney, and Little, 1975; Lilien, Kotler, and Moorthy, 1992):

$$S(t) = \frac{A(t)}{A(t) + Ac(t)}$$

[2] Note that the units of performance are in terms of a consumer performance index which can be assessed at the various stages of the design process.

where A(t) and Ac(t) are attraction levels of the firm's and competitors' products in the marketplace, respectively. The attraction of the new product, A(t), is assumed to increase linearly with the mean product performance and decrease linearly with the variance of the product performance. That is:

$$A(t) = \mu(t) - \rho V(t)$$

where ρ is a proportionality constant. Fourth, the unit cost of a product is increasing and linear in mean and variance of the product performance at the time of product launch (c.f. Porteus, 1986; Moorthy and Png, 1992).

$$c\,(\mu(t), V(t)) = \theta_m\, \mu(t) + \theta_v V(t)$$

where θ_m and θ_v are marginal costs of the mean and variance of product performance.

Revenue from new product sales can only be realized in the Marketing phase, $[T_p, T]$. During $[0, T_p)$ we allow for a revenue stream from an existing product. We assume that upon introduction, the new product makes the old product completely obsolete (e.g. the 1992 Toyota Camry makes the 1991 model completely obsolete and the latest version of a software program often makes its predecessor completely obsolete). The firm's cumulative profit is determined as follows,

$$T_\pi(t) = TR(t) - TC(t)$$

where TR(t) and TC(t) are net revenues and costs at time t, respectively. The net revenues function is given by

$$TR(t) = \int_0^t [\, p(s) - c(\mu(s), V(s))\,]\, M(s)\, S(A(s))\, ds$$

where $c(\mu(.), V(.))$ = unit production cost determined at the beginning of the Marketing stage and M(.) is the total market size at time t[3]. We note that the production cost includes direct costs (material, labor), indirect and overhead costs, exclusive of the costs of product and process design and marketing effort. The total cost function is given by:

$$TC(t) = TC(T_{i-1}) + C_i K_i + \int_{T_{i-1}}^t W_i(s)\, L_i(s)\, ds$$

This total cost function assumes that labor costs at stage i are charged at wage rate $W_i(t)$, at time t. Capital inputs are lumpy and are charged at unit cost C_i. It

---

[3] Note that before the new product is launched, $c(\mu(s), V(s)) = c_0$ which is the unit cost of the existing product. Also the attraction level of the firm's existing product in the marketplace is A_0.

is straightforward to discount all costs to time 0. For expository purposes, we ignore discounting.

The objective is to maximize the life-cycle (i.e. cumulative) profits $T_\pi(T)$[4], where T is the end of the window of opportunity available to the product. This objective is attained by optimally choosing labor resource and time allocation policies before the Marketing stage (i.e. $L_D(t)$, $L_P(t)$, τ_D, τ_P; for t in $[0, T_p)$). The optimization problem can be formulated as a multi-stage optimal control problem with labor intensity as the control variable and attraction and unit cost of the product as the state variables (see Cohen, Eliashberg, and Ho 1992b). This model structure is quite rich and can also be used to explore a number of interesting managerial issues. It also forms the basis for more extensive analyses involving multiple products, dynamic learning, and product generations. In the next section, we shall focus on the interpretations of the derived analytical results.

3. Testable Hypotheses

Based on few stationarity assumptions (see Cohen, Eliashberg, and Ho, 1992b), we obtain results 3.1 and 3.2. If, in addition, the size of the development team is fixed and there is no opportunity to enhance the perceived product performance in the Marketing stage, closed-form solutions for optimal time to market and product performance can be derived. These closed-form solutions lead to stronger results given in 3.3 to 3.5 (see Cohen, Eliashberg, and Ho 1992a for details).

3.1. Resource Allocation Across Developmental Stages. The optimal new product development policy is such that under certain circumstances it is optimal to focus development and time resources on the most productive developmental stage. At the individual project level, this policy implies that life-cycle profits can be increased via a focused resource strategy. This focused resource strategy appears to be practiced by Japanese automakers in the 70s and 80s when they channelled their development resources to improving the reliability of the car (i.e. Process stage). Tandy computer adopted the same strategy and subcontracted out its Process activities. Mansfield (1988) studied several industries in Japan and U.S. and reported that Japanese companies allocated their development resources unevenly across development stage. In particularly, an unusually large proportion of resources was allocated to Process

---

[4] Other objectives such as return on investment or a weighted average of profit and market share are possible under the framework.

activities. American companies, on the other hand, spread their resources more evenly across development stages.

At the firm level, this policy suggests that the firm should consider an increased specialization and less vertical integration of design activities. Firms will focus and capitalize on their design strengths and hire outside suppliers for their less efficient activities. These subcontract opportunities will make specialized design services viable and flourish. It is interesting to note that the popular notion of core competency is consistent with these findings and is operationalized here as the firm's productivity in delivering product performance per unit time.

3.2. Development Team Size. The structure of the optimal policy is such that it is best to have a fixed labor input throughout the development period. This result stems from the assumed diminishing return to labor investment in the performance improvement process. Since the diminishing return assumption is quite reasonable in practice, this result suggests that firms should strive for stability in the size of the development team throughout the development process. Having greatly varying team size over time is suboptimal.

3.3. Time to Market and Product Performance. The optimal time to launch a new product increases with both the margins of the existing and the new products. In addition, it increases with the total market size. These results suggest that in a high-margin or large size market, better life-cycle profits could be obtained if the firm would develop a "superior product" rather than adopt a "quick and dirty" strategy associated with rapid introduction of a lower performance product.

The time to market is an increasing concave function of the product life cycle. That is, the first derivative of time to market with respect to product life cycle decreases with the product life cycle. Thus, the tension to compress time to market will be even greater when product life cycle is shortened. As a consequence, time to market becomes an increasing fraction of the life cycle as the life cycle contracts.

3.4. Design Capability Hurdle. The design capability hurdle indicates the threshold level of design technology the firm must possess in order to profitably undertake the development project. The model predicts that the threshold will be higher if the total product performance in the market is high; will be lower if the total market size is high, the product life cycle is longer, and the competitor has most of the market shares. The result that design capability hurdle is lower if the firm has a lower market share is interesting because it suggests that an incumbent has a higher design capability hurdle than a potential entrant in undertaking a new product development project.

3.5. Cannibalization. The model predicts that it is optimal to delay product launch if the new product cannibalizes an existing product. The amount of delay is larger if the product life cycle is longer and if the performance level of the existing product is high; and is smaller if the firm has an efficient design technology.

3.6. Exploitation of Improved Design Technology. The model suggests that different product developers should invest and exploit technology which improves the design process (e.g. CAD/CAM) differently. More efficient product developers should use the improved technology to both increase product performance and decrease time to market simultaneously. Less efficient firms, on the other hand, should only use the improved technology to enhance the product performance. It is not optimal for the less efficient firms to use the improved technology to shorten time to market. This result challenges the conventional wisdom that it is always optimal to use improved design technology to compress development cycle.

4. Evaluation of Industry Practices

Our normative framework represents the best possible policy to which several reported industry practices are to be compared. These industry practices simplify the product development process by either imposing one or more constraints on the optimization problem or by ignoring some interactions among the success factors. An evaluation of these industry practices based on a preliminary numerical simulation experiment[5] is conducted. The results are reported below:

4.1. Traditional Approach. The traditional approach ignores the interaction between unit cost of the product and the product performance. In terms of optimal control terminology, the kinematic equation for the unit product cost is ignored in the optimization problem. The approach starts with an initial estimate for the unit product cost and checks the resulting unit product cost at

[5] The experiment manipulated four exogenous parameters: 1) the length of the window of opportunity, T, 2) total market size, M, 3) price, p, and 4) competitive product attraction, Ac, Each of these parameters has three levels. Thus, the numerical experiment consists of 81 examples.

the conclusion of the design process. If the resulting cost is too high, then the firm must go back and repeat the design process (see Figure 3)[6].

Our numerical simulation results show that the approach often results in a grossly suboptimal new product development policy. In particular, the approach leads to over-sized development teams, too ambitious product change levels, inflated time to market, and resulting unit product costs which are too high. These results suggest that the interaction between unit product cost and product performance is highly relevant for the management of the new product development process.

4.2. Target Costing Approach. It has been suggested that the secret weapon for Japanese success lies in its use of target costing approach to design (Fortune, 1991). The approach is similar to the optimal policies except that there is a constraint imposed on the unit cost of the product (See Figure 3). Our analysis predicts that if the target cost is not too far away from the optimal unit product cost, then the deviation from the optimal solution is small. This result is interesting because it implies that target costing approach might be a good approximation for best policy if the firm has a reasonable estimate of what the optimal unit product cost is. It can be shown that the size of the development team, product performance, and time to market are lower (higher) than the corresponding optimal value of the best practice if the target cost is lower (higher) than its optimal value.

Compared to the traditional approach, the target costing approach has a shorter time to market even if the target cost is set much higher than its optimal value. The approach also has a smaller development team size and makes a less revolutionary product innovation than the traditional approach. These results are consistent with the empirical observation that American companies which adopt the traditional approach to design often introduce more revolutionary products but at a lower frequency. Compared to the traditional approach, the target costing approach makes significantly higher profits.

---

[6] In the numerical simulation experiment, the resulting unit product cost is always less than the price and we assume that no iteration is necessary. In practice, the firm might have a target unit cost. If the resulting unit cost is higher than the target cost, the firm will go back and repeat the design process. Our framework, in its present form, is open-loop and does not have a mechanism to capture this iterative process. The framework is currently being extended so that it can capture this dynamic aspect of the traditional design process.

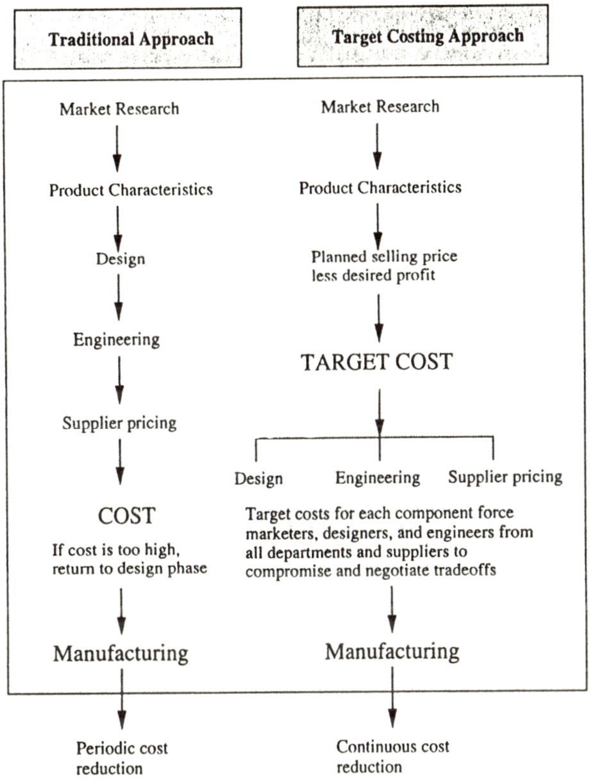

Figure 3: Two Design Approaches for New Product Development (Fortune, 1990)

4.3. Target Performance Approach. As the name implies, the approach imposes a constraint on the product performance target to simplify the product development process. The approach can be used by followers in a product market to catch up with the leader. Some innovators use this design approach to ensure that their new products are always at a fixed percentage superior to the best product in the market. In other cases, firms are designing to meet a current or anticipated industry standard. Like the target costing approach, the approach considers all interactions among the success factors.

The framework indicates that setting a higher than optimal product performance level target leads to longer time to market, larger development team size, and higher unit product cost. A lower than optimal product performance level target has reverse effects. In both cases, the loss in life-cycle profits are less severe than the traditional approach if the target performance is reasonably close to its optimal value.

4.4. Target Timing Approach. The target timing approach is particularly relevant in industries where there are natural product introduction times (model year and season). Examples include automobiles (beginning of the year), toys (Christmas season), and wearing apparel (fashion season). The approach considers the interactions among all success factors but imposes a constraint on the time to market.

The framework predicts that if the time to market is set higher than the optimal time to market, the size of the development team is smaller, the product performance is higher, and the unit product cost is higher than their corresponding optimal values. Imposing a lower than optimal time to market has reverse effects. Again, the loss in life-cycle profits is as not severe as that of the traditional approach.

Table 1 provides a summary of the results. The optimal values of development team size, time to market, product performance, and unit product cost of each design practice are compared to those of the best practice under two scenarios. Timing, performance or unit cost targets (or in the traditional approach initial cost estimates) which are lower or higher than the optimal values of best policy are chosen. The numerical values are percentage deviations from the optimal values of the best policy. It is interesting to note that the loss in profits resulting from imposing a constraint on time to market, product performance level target, or unit product cost is not as severe as that of ignoring the interactions between the success factors. Thus, firms should invest in acquiring know-how of the underlying relationships among the success factors by analyzing their design capabilities and the tradeoffs inherent in the process. The target costing, performance or timing approaches are good alternatives to the best practice (optimal solution) since they simplify the new product development process considerably with a small profit penalty (as long as these targets are in the neighborhood of the optimal value).

4.5. Minimize Break-even Time. We can also analyze the impact of minimizing break-even time (BET). Hewlett Packard has announced that one of its corporate strategies for the 90s is to reduce product break-even time by one half of all its new product (House and Price, 1991). Our framework shows

Design Practice	Development Team Size	Time to Market	Product Performance	Unit Product Cost	Profits
1. Traditional					
0.95 c^*	+48.3%	+51.3%	+84.5%	+16.0%	- 25.4%
1.05 c^*	+23.3%	+56.1%	+73.5%	+13.9%	-19.8%
2. Target Costing					
0.95 c^*	- 7.2%	-24.1%	-26.9%	- 5.0%	- 4.1%
1.05 c^*	+ 7.1%	+22.7%	+26.9%	+ 5.0%	- 3.3%
3. Target Performance					
0.80 A^*	- 3.8%	-18.4%	-20.0%	- 3.7%	- 1.9%
1.20 A^*	+ 6.5%	+16.3%	+20.0%	+ 3.7%	- 2.1%
4. Target Timing					
0.80 T_p^*	+42.5%	-20.0%	- 4.5%	- 0.8%	- 0.5%
1.20 T_p^*	-25.7%	+20.0%	+ 3.4%	+ 0.7%	- 0.4%

Note: x^* is the optimal value for variable x of the best policy generated by the model framework

Table 1: An Evaluation of Four Industry Practices vis-a-vis the Best Policy Generated by the Model Framework

that minimizing BET often leads to premature product introductions that exhibit conservative bias in product performance change. Thus, BET alone might be too simplistic a metric for managing new product development process (Cohen, Eliashberg, and Ho, 1992a). It is interesting to note, however, that depending on the current value of BET, the decision to achieve BET/2 could move a company closer to the optimal value for BET.

5. Conclusions

In this paper, we develop an analytical framework for generating managerial insights for new product strategy. The insights are stated as testable hypotheses so that they can be subject to empirical scrutiny. We also apply the model framework to evaluate several industry practices and benchmark each practice against the best policy generated by the framework. Each practice is a restricted case of the optimal solution generated by the model framework and thus is sub-optimal if all practices have identical costs of implementation.

REFERENCES

Bell, D., R. Keeney, and J. Little, "A Market Share Theorem," Journal of Marketing Research, 12:136-141.

Crosby, P. B., Quality is Free, McGraw Hill, New York, 1979.

Cohen, M., J. Eliashberg, and T. H. Ho, "New Product Development: Performance, Timing, and the Marketing-Manufacturing Interface," Decision Sciences Working Paper, The Wharton School, University of Pennsylvania, 1992a.

Cohen, M., J. Eliashberg, and T. H. Ho, "New Product Design Strategy: Evaluating Industry Practices," Decision Sciences Working Paper, The Wharton School, University of Pennsylvania, 1992b.

Cohen, M and Dougherty, D. "From Concurrent Engineering to Concurrent Management: A Framework for Being an Innovative Organization," Decision Sciences Working Paper, The Wharton School, University of Pennsylvania, 1992.

Fortune, "Japan 's Smart Secret Weapon," August 12, 1991.

Graves, S., "The Time-Cost Tradeoff in Research and Development: A Review," Engineering Costs and Production Economics, 16(1989): 1-9, 1989.

Grossman, G. and Helpman, E., "Quality Ladders in the Theory of Growth," Review of Economic Studies, 58:43-61, 1991

Juran, J. M., and F. M. Grana, Jr., Quality Planning and Analysis, 2nd ed., McGraw Hill, New York, 1980.

Karmarkar, U. and R. Pitbladdo, "Quality, Class, and Competition," Working Paper, William E. Simon Graduate School of Business Administration, University of Rochester, 1992.

Lilien, G., P. Kotler, and K. S. Moorthy, Marketing Models, Prentice Hall, Englewood Cliffs, New Jersey, 1992.

Mansfield, E., "The Speed and Cost of Industrial Innovation in Japan and United States: External vs. Internal Technology," Management Science, 34(10):1158-1170, 1988.

Moorthy, S., and I. P. Png, "Market Segmentation, Cannibalization, and Timing of Product Introductions," Management Science, 38(3):345-359, 1992.

Mussa, M., and S. Rosen, "Monopoly and Product Quality," Journal of Economic Theory, 18, 301-317, 1978.

Porteus, E., "Optimal Lot Sizing, Process Quality Improvement and Setup Cost Reduction," Operations Research, 34, 137-144, 1986.

Smith, P. G. , and D. G. Reinertsen, Developing New Products in Half the Time, Van Nostrand Reinhold Book, 1991.

4

FROM PROJECT TO PROCESS MANAGEMENT IN ENGINEERING: MANAGERIAL AND METHODOLOGICAL CHALLENGES

Paul S. Adler*
Department of Management and Organization
School of Business Administration
University of Southern California
Los Angeles CA 90089-1421

Avi Mandelbaum
Faculty of Industrial Engineering and Management
Technion
Haifa, Israel, 32000

Viên Nguyen
E53-353
Sloan School of Management
Massachusetts Institute of Technology
Cambridge MA 02139

and

Elizabeth Schwerer
Graduate School of Business
Stanford University
Stanford CA 94305-5015

Acknowledgments: This study would not have been possible without the generous cooperation of the company we have called Chemicals, Inc. We owe thanks in particular to the manager and staff of its Plastics Division. This project has also benefited from the advice and support of Chuck Holloway and Mike Harrison. Funding from the Stanford Integrated Manufacturing Association is gratefully acknowledged. Elizabeth Schwerer was supported by a National Science Foundation graduate fellowship.

---

* The order of the authors' names is alphabetical

Abstract: Engineering product development efforts have often been studied as though they consisted of independent and idiosyncratic projects. In reality, however, many product development organizations do not rely on fully-dedicated teams, so their projects suffer delays when resources have to support more than one project concurrently. Moreover, in many product development organizations, projects often exhibit important similarities with previous projects. While PERT models allow the characterizations of independent and idiosyncratic projects, a more realistic model of development organizations would therefore represent them as stochastic processing networks in which engineering resources are "workstations" and projects are "jobs" that flow among the workstations. This paper describes the managerial issues involved in applying such a process approach to product development, and the methodological issues in modeling development time with this approach. We identify these issues through a field-based research project.

1. Introduction

The motivation for the research reported here was the growing competitive importance of product development time. As argued by business observers such as Blackburn (1991), Clark and Fujimoto (1989a), Stalk and Hout (1990), and Wheelwright and Clark (1992a, 1992b), the intensification of global competition has created enormous pressures on firms to accelerate the process of developing and launching new products.

Our research was premised on two hypotheses concerning the product development process. First, while product development projects are often viewed as collections of unique activities, we believe that in reality different projects within a given organization often exhibit substantial similarity in the overall flow of constituent activities. Second, while most of the planning tools available to managers assume that projects are independent clusters of activities, we believe that in reality many organizations must manage concurrent projects that place competing demands on shared human and technical resources.

The conjunction of these two hypotheses suggests a proposition: a process view — that is, a view of the development process as a more or less repetitive "design production process" — can provide a framework for better understanding product development time in organizations with multiple concurrent non-unique projects that utilize shared resources. In particular, such a process view would allow us to estimate the likely delays of projects awaiting resources and to identify otherwise invisible opportunities to accelerate time-to-market.

This proposition can be restated in organization-theoretic terms: the degree of "structure" or "programmability" of engineering tasks is in part "enacted" rather than exogenously given (see Weick, 1977, on the concept of enactment). Whereas the naive researcher might take at face value an informant's description of engineering tasks as much less structured and programmable than manufacturing tasks, closer analysis might reveal the possibility of "reenactment" both in the weaker sense of the term — tasks that are interpreted as unstructured might reasonably be interpreted differently — and in the stronger sense — the real tasks can be reshaped to make them more structured.

Several previous studies have suggested that product development time could be significantly shortened by the adoption of a process approach (Hayes, Wheelwright and Clark, 1988; Schoenberger, 1986; Alexander, 1990), but they have not tested the idea by attempting to model realistically a sample firm's product development activities. The project described here sought to build such a model. Two other papers (Adler et al., 1993a and 1993b) describe the operational results of this project. The research process itself also revealed important lessons concerning the technical methodological challenges of process modeling in engineering contexts. Under many of these methodological issues lie organizational challenges created by a shift from a project management to a process management viewpoint. These managerial and methodological issues are the focus of the present paper.

With the aim of better identifying and understanding these issues, this paper recounts the method by which we constructed a process model for a sample engineering organization and the challenges we encountered in this effort. The following section summarizes the formal characteristics of process models of the kind we envisaged. We then highlight several issues that we considered in choosing a research site and describe the organization that eventually became our host. The next sections discuss the issues we encountered in operationalizing and measuring the process model variables. A concluding section highlights the lessons learned.

2. The Conceptual Framework for Process Modeling

The most commonly used tools for predicting product development times are descendants of PERT (Project Evaluation and Review Technique) and CPM (Critical Path Method) (Dean, 1985). But these techniques fail to account for two key phenomena that can slow project progress.

First, they depict an idealized flow of project activities in which activity times are relatively predictable and first attempts always succeed. By contrast, many product development organizations face uncertainties in activity times and in number of activity repetitions. For example, proposed designs are often tested and iteratively redesigned and retested until specifications are met. Moreover, the duration of prototype testing cycles may vary from one iteration to the next.

Second, typical PERT/CPM analyses, if they acknowledge resource limitations at all, assume that resources are dedicated to a single project at a time. Thus they fail to account for the congestion effects arising from contention for resources among concurrent projects. Although there are extensions of PERT/CPM techniques that acknowledge constraints on resources, they operate in a *static* environment where no new projects are introduced over time.

We posit that the product development organization can be viewed as a system whose activities can be described in terms of processes. This approach differs from previous studies of product development in that it focuses on the management of resources supporting multiple concurrent projects instead of focusing on the management of individual projects.

Accordingly, the research reported here sought to model the product development organization as a stochastic processing network. Exhibit 1 depicts such a network. For our purpose, a stochastic processing network consists of a collection of "workstations" or "resources," each of which is composed of one or more identical "servers" working in parallel. A workstation corresponds to a pool of employees, typically with the same job title, who perform the same functions interchangeably. For example, two of the resources in our model are "process engineers" (a pool of two employees) and "manufacturing engineers" (a pool of five). The servers are the technicians or engineers who make up the pool. The organization processes projects, or "jobs," which consist of collections of tasks (such as "Product Prototyping" and "Product Testing") that are performed by specified resources in specified sequences. Certain tasks can be carried out in parallel while others must be performed sequentially. When several tasks may begin processing at the same time, we refer to the phenomenon as a "fork"; when a task may not begin until several other tasks have been completed, we call it a "join." The time required to complete a task is called its "processing time," or "activity time," and the intervals between the starts of new projects are called "interarrival times."

The queue at each workstation (represented by shaded boxes in Exhibit 1) corresponds to the "in-box" of the resource. A complete description of the model must specify the service discipline at each station — that is, the rule by which the server chooses the next task from the queue. We implement our basic model using the "round-robin" discipline within priority classes. In this discipline, a free server takes the next top priority task in the queue and works on it for a pre-specified length of time. If he completes the task within the time period, the corresponding project moves on to a successor task, and the server continues with the next top priority task in the queue. Otherwise, the task returns to the end of the queue, its remaining processing time is updated to reflect the last round of service, and it waits until its next access to the server. When the queue is empty of top priority work, the station serves the second priority tasks in the same manner. Clearly there are other possible choices for service discipline, including first-in-first-out, last-in-first-out, or project with the earliest due-date first; in Adler et al. (1993b) we explore the impact on project completion time of switching to a first-in-first-out discipline.

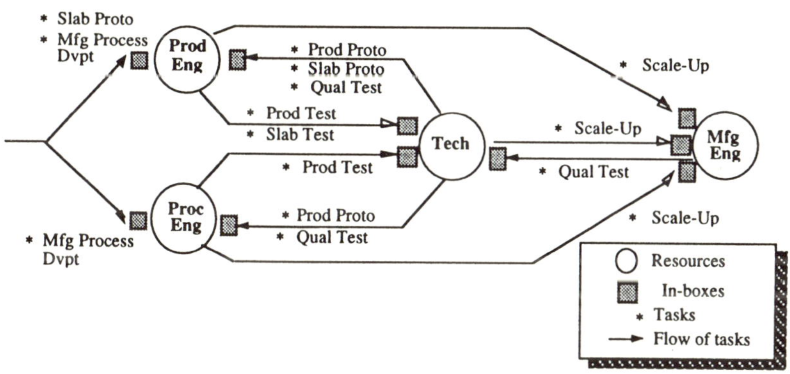

Exhibit 1: Processing Network Representation

We can use PERT-style diagrams to illustrate constraints on the order in which tasks are executed. For example, Exhibit 2 is the PERT diagram associated with the processing network depicted in Exhibit 1 — these are parts of the product development process that will be described in the following section. Each job consists of seven activities. Activities "Manufacturing Process Development" and "Slab Prototype" can be performed in parallel (they represent a fork) and "Scale-Up" begins when activities "Product Testing" and "Manufacturing Process Development" are both completed (a join).

The ease with which tasks fork is one of the features that distinguishes engineering from manufacturing tasks. Whereas a drawing and a CAD file can be reproduced to allow different engineers to work in parallel, manufacturing processes typically do not allow for a component or subassembly to fork into different parallel processes. Process models for manufacturing environments must allow for joins (as components and subassemblies come together), but they do not in general need to confront the greater modeling complexity of forking processes.

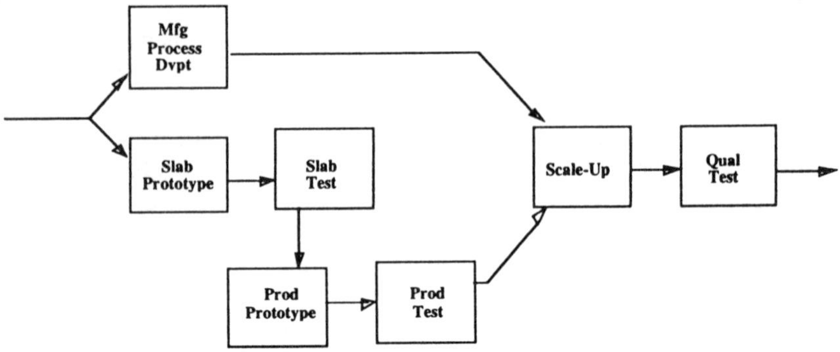

Exhibit 2: Traditional PERT Representation

Alongside the technical difficulty of modeling forks there is a real management challenge: engineering organizations experience considerable freedom as to where to fork. In practice, there is considerable freedom in deciding the degree of parallelism that should be built into the engineering task flow. The technical pros and cons of forking at any given step are not always obvious and in some cases a range of possible workflow design options may be equally effective. This insight helps us understand why organizational politics and culture often weigh so heavily in shaping the roles that different firms assign to different engineering groups (e.g. design engineering and manufacturing engineering) in the product development process. Process modeling efforts such as the one we report here must rely on engineering assessments of the set of feasible alternative workflows, but the models can help compare the congestion effects associated with each alternative.

The processing network is stochastic because interarrival times, processing times, and precedence requirements may be subject to statistical variability. Although these quantities are random, they can be characterized via probability distributions. Projects are said to be of the same "type" if their individual precedence requirements, processing times, and interarrival times can be characterized by the same set of probability distributions.

Underlying this framework are several assumptions, and as the history of our project revealed, these assumptions have significant methodological and managerial corollaries. First, we must assume that the organization's tasks and technologies are stable over time and that projects can be characterized by sets of probability distributions. We must further assume that there is a small number of identifiable types of projects and that projects within each type are similar in the sense that differences between their realizations can be attributed to stochastic variability.

3. The Research Site

In searching for a host, we found that the assumptions underlying process modeling and mentioned in the previous section did not fit all engineering environments. Since our model requires significant similarity among projects, it appeared to be ill-suited for groups involved in the basic research end of the R&D spectrum, where it would seem that projects are more idiosyncratic (although further research might usefully challenge this assessment). Similarly, our assumption of stable probability distributions seems inappropriate for groups whose technologies, products, or organizations change rapidly. Furthermore, the assumption that resources divide their time among competing projects is critical to our focus on the delay created by the contention for resources; in organizations that rely on dedicated teams – as many software organizations do – problems such as underutilized resources may arise, but resource contention typically does not.

We eventually found a host in what we will call the Plastics Division of Chemicals Inc. This division offered several advantages as a research site. First, both divisional and corporate management were interested in understanding how they could accelerate their product development cycle, so they were eager to help us in our research. The Plastics Division had recently lost several possible contracts because their principal competitor, a large Japanese firm, had been significantly faster in developing new products. Management support was crucial to the success of our project because of the time and effort involved in helping us collect data. In addition, for several years the Product Development manager had been collecting time cards from his staff, and these could be retrieved for our use.

The Plastics Division at Chemicals Inc. made plastic parts. It accounted for some 7% of Chemical's total domestic sales. Historically, the Plastics Division had primarily sold custom-designed products for the aerospace and defense industries. With the slowdown in defense contracts in recent years, the Plastics Division was shifting its focus to the automobile industry, an effort that coincided with that industry's increased use of plastics. In moving away from defense and into a commercial market, the Plastics Division was increasingly concerned with cost and timeliness issues and with high-volume production.

The staff of the technical department in the Plastics Division consisted of engineers and technicians divided into functional groups specializing in product design, process design, and applications. A technical services group supported these engineering groups by helping to make and test product prototypes. Finally, manufacturing engineers, product managers, salespeople, specifications specialists, and other staff members all made critical contributions to development projects. These resources constitute distinct workstations in our network.

Although the group considered its principal mandate to be the development of new products, it also handled "reformulations" — projects to replace the materials in existing products — and it supported products on the market. New product development and support efforts were typically triggered by customer interest; reformulations arose either when a vendor discontinued a constituent material or when a better material became available. Reformulation projects usually had little urgency: the plant might have several months' supply of the current material, and the potential cost savings from using a different material was typically small. Only rarely did circumstances force reformulation projects into top priority.

Since the overall need for product reformulations was sensitive to the evolution of the materials market, the product development group's work tended to oscillate between new products and reformulations over periods of several years. Support activities, on the other hand, remained at a fairly stable level. When the Plastics Division restructured its engineering groups a few years before our study, the new manager of product development discovered that he needed to "clean up" operations and he therefore focused some 70% of the group's efforts on reformulations. The more recent Plastics Division strategy of targeting the automobile industry required a new line of products, so efforts shifted back to new product development. At the time of our project, the Plastics Division wanted to strengthen its competitive position through lower costs, so the product development group expected to see an increasing proportion of reformulation projects in future years. (An important feature of our model is its ability to reflect the changing mix of work.)

Management assigned formal priorities to projects to help resources allocate their time. Typically, projects involving the development of new products were given priority 1 (the higher priority) whereas most reformulation projects were treated as priority 2. This priority system also affected project trajectories by communicating to resources outside the technical department how they should treat projects. Managers and engineers often expressed exasperation over the long delays that priority 2 projects suffered while waiting for attention from product managers or the manufacturing plant. If the priority reflected the true business importance of the project, then such delays might seem inevitable and even appropriate. Nevertheless, one project we studied had spread two person-months of work over 2.5 years, raising questions about inefficiencies due to mental set-up time, the opportunity cost of delay in getting the product to market, and the toll of prolonged management distraction. One purpose of our project was to quantify the delays arising from various management policies so that these costs could be evaluated more explicitly.

Once we began working with the Plastics Division to collect data, we realized that the host organization offered another important advantage. The group had recently defined a standard, five-phase procedure for product development. This procedure specified the activities that needed to take place in each phase and the conditions the project had to satisfy before moving on to the next phase. This phase system proved essential to our data collection effort by providing a standard nomenclature and a common understanding of the specific tasks constituting the product development process.

4. Operationalizing the Framework

In this section, we describe how we operationalized our model's components. The development of each construct was associated with methodological challenges, many of which in turn pointed to important managerial issues.

Project types. To characterize statistically the projects at the Plastics Division, we asked our informants to identify major categories of projects according to the similarity of the projects' activity histories. This is generally the first step in systematizing any process, but unlike the explicit and formalized process plans in manufacturing operations, process plans in the Plastics Division, as in many product development organizations, were largely tacit. When asked to categorize types of projects using their activity flow configuration as a criterion, the Plastics Division's employees responded initially with rather ill-formed and inconsistent suggestions. Our question forced them to impose a taxonomy on their work portfolio, and it took several meetings to define an acceptable clustering.

Our categorization criteria for project types required that each type occupy a substantial amount of resource time and that each have some structural features distinguishing it from other types. Some dialogue with our Plastics Division informants was needed to operationalize this general specification, and in the process, our research team was forced to clarify our own understanding of engineering process modeling. Through this dialogue, we identified three main features of a project: the assigned priority, the characteristics of the final product, and the nature of the development effort as reflected in the constituent activities and their precedence relationships. The assigned priority affected the service discipline applied to the project. The characteristics of the product determined the technical requirements and hence influenced the development activities. Finally, the nature of the development effort determined the required resources and activity sequence.

Following the suggestion of our key contact at the Plastics Division, we focused our study on a family of products we will call plastic parts. This product family accounted for over 80% of the engineering organization's time. The other family was made up of more idiosyncratic projects developing more complex systems. As in the broader class of all products, plastic parts work included both new product developments and reformulations, and these projects could be assigned either priority 1 or priority 2.

In summary, our model consisted of two project types — new products and reformulations — and two priority levels for each. Our decision to isolate these four types of projects accomplished two goals. First, it allowed us to explore how various project characteristics affected the product development cycle. Second, with these two categories of projects, we managed to capture the bulk of the Plastics Division's time. In combination with support activities and administrative duties (which the group also treated as priority 1 work), they accounted for virtually all of the group's time.

Resources. The core resources were the product and process engineers and technicians who dedicated their time to product development; but this "development group" also relied on several other resource pools. For example, they ordered materials from other divisions, ran prototypes in the manufacturing plant, requested tests from the technical services group, sought marketing and sales advice about the concerns of the lead customer, consulted with the specifications group about possible legal issues, and relied on product management to coordinate and promote these activities.

Our data gathering effort revealed wide differences in the quality of information available to us about these resources. Considering these information discrepancies, we distinguished resources that satisfied the following two criteria: the group had a potentially significant impact on the project process, and its activities were adequately documented. Those groups that potentially affected the rate of project completion but that did not satisfy the latter requirement were combined into a group called "Miscellaneous" which includes the following functions: Sales, Finance, Specifications, Logistics, and Quality Assurance.

An interesting methodological issue emerged in how to reflect the fact that resources may be both specialized and adaptable. Engineers and technicians, unlike their counterparts in manufacturing environments — machines and equipment — are capable of handling a wide variety of tasks beyond their formal functions in the organization. For example, during busy periods at the Plastics Division, the development group might turn some of its work over to engineers with nominally different functions, such as manufacturing or applications engineers (who normally deal with factory and customer implementation issues respectively). Since specialization is often a matter of organizational choice rather than technical capability, it is up to the modeler to decide whether to characterize this specialization as fixed versus variable. We discuss this issue in greater depth in the next section.

We identified nine resources. According to management estimates of resource availability (number of work hours per week per server within each resource), average work weeks varied from 40 to 55 hours depending on the resource. However, our simulation experiments (reported in Adler 1993b) later led us to recognize the need to allow for "crunch" periods with significantly higher levels of overtime.

Tasks. To find a partition of the product development activities at the Plastics Division, we turned to the phase procedure mentioned earlier. This procedure described five sequential phases of the development process, specifying a standard protocol for resolving the key issues in each phase. Phase 1 ("Concept/Feasibility") was characterized by the intensive involvement of a few marketing and product development people who simultaneously explored technical, manufacturing and market feasibility. In Phase 2 ("Project Plan/Team") a full team was assembled, and a project plan was drafted. Phase 3 ("Product Development") signaled the project's peak effort, as the team worked out the technical, legal, and marketing issues in detail. It was here that the development group expected to face the project's critical challenges. Phase 4 ("Manufacturing Standardization/Launch") marked the transfer of the project from the development labs to full-scale manufacturing. It included a concentrated effort to eliminate any remaining technical wrinkles, and it closed with the product launch. Finally, Phase 5 ("Continuous Improvement") represented ongoing refinements while the product was on the market. Each phase consisted of approximately a dozen issues to be resolved.

Although the phase procedure provided a detailed characterization of the relevant issues, it did not identify the discrete tasks that were required to resolve these issues. Something that appeared as a simple item to check off in a phase review meeting might require several distinct tasks by independent resources, or it might represent an issue that would be addressed several times as a project progressed. In the words of one of our contacts, "the phases are defined by objectives, not by activities." When a modeler defines a cluster of possible project steps as an activity, she assumes that this cluster can be viewed as a single entity across a spectrum of projects. However, the degree to which an employee aggregates activities in his description of a process depends on both his knowledge of the details of others' work and his knowledge of the function of his own work in the entire project. Thus any model of an engineering organization will depend on individual judgment — and this to a greater degree than in manufacturing. Our informants found it useful that our study forced them to characterize their tasks more explicitly than they had done before.

We considered several trial representations before arriving at a collection of activities that the Plastics development group found satisfactory. To simplify the data collection and the analysis, we decided to focus on the first four phases (through Manufacturing Standardization/Launch), aggregating Phases 1, 2, and 4 into one "activity" each and specifying Phase 3 (Product Development) in greater detail. We chose to highlight Phase 3 of the development process because it contained the bulk of project work and because it illustrated some interesting network features which distinguish product development from manufacturing, namely, forking, iterative looping among activities, continual transmission of information between steps, and resources who potentially juggle several activities at once. Phase 3 later proved to be a felicitous choice for the additional reason that it marked the time of crispest activity definition.

Our model of the development process for new product projects consists of 18 activities. Each of these activities may require attention from several resources. We therefore developed an "activities-resources map" to display the involvement of resources in each of the 18 activities (see Exhibit 3). A "task" corresponds to an activity/resource pair (e.g. Review Patent/Product Engineer). Thus while each task is performed by exactly one resource, each resource can be responsible for several tasks.

Reformulation projects tended to be smaller projects than new product development efforts, requiring fewer resources and fewer person-hours to complete. Often they skipped Phases 1 and 2 entirely and began with Phase 3.

Precedence Constraints. Simultaneously with our effort to identify tasks, we asked our key contact at the Plastics Division to translate the phase system into PERT-like diagrams illustrating the flow of activities. We wanted to find abstract representations of projects that our informants could regard as both meaningful and realistic, and we needed many rounds of discussion to identify a satisfactory flow diagram.

The activity flow diagram presented in Exhibit 4 reflects the resulting model of the process flow. Activities are shown in boxes and arrows indicate precedence among activities; if several resources were involved in an activity, we assume that they could execute their tasks in parallel. An "ideal" project would have only forward arrows: that is, a downstream activity would never be followed by an upstream activity. Exhibit 4 represents the following process flow: a new project begins at Phase 1 and then proceeds to Phase 2. Its completion of Phase 2 triggers the start of several (possibly) simultaneous Phase 3 activities. Product engineers and technicians begin activities in the prototyping cycle: making and testing material samples ("slabs") to refine the material formulation, making and testing product prototypes in the lab to explore product geometries and to study the material's behavior, and finally making product prototypes in the plant to uncover manufacturability issues. Product engineers, process engineers, and technicians simultaneously develop the manufacturing process, getting information from the product engineers about special product requirements and sharing their own cost and feasibility results.

	Prod Engr	*Proc Engr*	*Prod Tech*	*Proc Tech*	*Tech Svcs*	*Prod Mgt*	*Misc*	*Mfg Engr*	*Appl Engr*
Phase 1	X		X			X	X		X
Phase 2	X	X	X	X	X	X	X	X	X
Rev Patent	X		X						X
Mfg Proc Dvpt	X	X	X	X			X	X	
Mkt Position						X	X		
Make Slabs	X		X						
Test Slabs	X		X						
Make Product	X	X	X	X				X	
Test Product	X		X						X
Make Prod-Mfg	X	X							
Test Prod-Mfg	X		X						X
Sales Strat	X					X	X		X
Lead Customer	X	X				X			X
Product Specs	X	X			X	X	X		
Field Trials	X	X				X			X
Agency Specs	X						X		
Qual Testing	X	X	X	X	X	X	X	X	X
Phase 4	X	X	X	X		X	X	X	X

Exhibit 3: Activities–Resources Map for New Product Projects

At the same time, people in sales and product management begin the activities shown in the lower-left corner of the flow chart. These activities initially have little impact on the technical side of product development, but ultimately the two groups negotiate in the course of the product specifications effort. When a final product and set of specifications are defined, technical services performs a comprehensive set of qualification tests to ensure that the product meets these specifications. If all goes well, the new product proceeds on to field trials, to the phase 3 (design) review, and finally to the full manufacturing scale-up and launch of Phase 4.

The proliferation of reverse arrows in the flow chart illustrates the phase system's purpose of systematizing checks. Since these "activities" were really objectives, each of them could be regarded as a decision box indicating whether earlier issues needed to be revisited before the project moved on. For the sake of simplicity in our model, we included only those iterations that our Plastics Division informants believed occurred with significant frequency.

The process flow diagram for reformulation projects was simpler. The process typically bypassed Phases 1 and 2, and Phase 3 contained fewer activities, although sometimes there were just as many iterations between slab and product prototyping with the introduction of a new material. Our model reflected reformulation project's relative simplicity.

5. The Challenges of Data Collection

The preceding section discussed the key variables of our model and how we arrived at operational definitions for them. In order to derive quantitative conclusions from our model, we needed statistical data characterizing its components. For each job type, we required data on the frequency of new job starts (interarrival times), the tasks involved and the order in which they were executed (precedence requirements), and the time to complete each task (service times). Because we accounted explicitly for time spent on support and administrative duties, we also needed data on the time that each resource devoted to these activities. Moreover, our model required the entire distribution of these quantities and not only their means.

Lacking an appropriate data base, we began this data collection effort by asking our informants at the Plastics Division to estimate averages for a recent new product project — Project X — which had enjoyed high visibility in the group and had been well documented. We intended this preliminary exercise to provide a baseline from which we could ask our informants to generalize to probability distributions for the entire portfolio of new product projects.

We collected our data in three half-day workshops with a group of Plastics Division employees representing most of the resource pools. These meetings proved to be rich data-gathering efforts because they raised fundamental issues about our model — for example, would our informants find it meaningful to assign probability distributions to events that they viewed as essentially unique? We had extensive conversations about some of our simplifying assumptions, sometimes resulting in further aggregation of activities in the flow diagram,

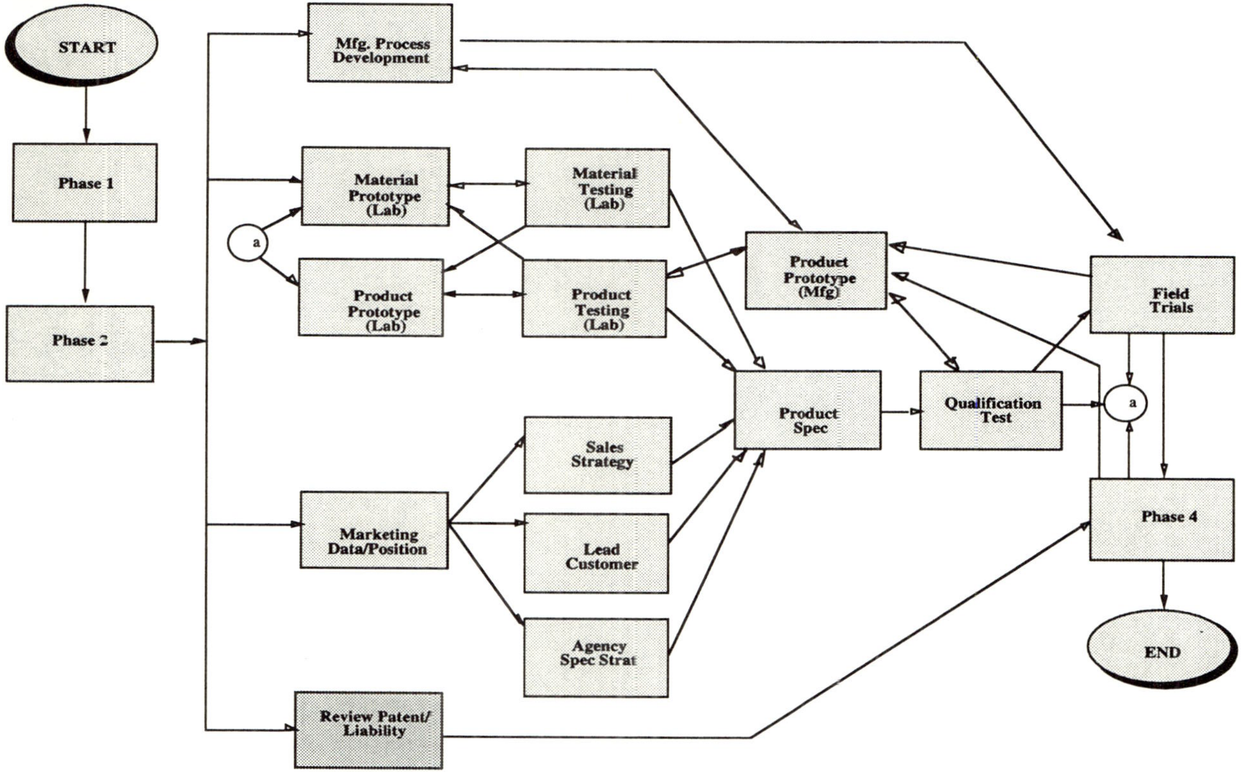

Exhibit 4: Process Flow Diagram – New Products
Note: A feedback arrow links the circles marked "a."

and sometimes highlighting complicating issues that might affect the validity of our results. For example, we learned that the boundaries between phases were not hermetic: some phase 3 activities actually began in phases 1 and 2, and some continued into phases 4 and 5. Our sources emphasized a similar phenomenon at the level of activities within phase 3: many pairs of nominally sequential activities actually overlapped or involved significant passing of information back and forth. (Although we have not included this continual release of information in our model, we think it is a feature of product development that should be explored in future research since it involves a critical trade-off between the value of early communication and the risks of basing work on incomplete information.)

A further issue we considered in these workshops was how to deal with project delays caused by resources outside the division (and hence beyond divisional control). For example, the product development group relied on manufacturing to run "scaled-up" production of prototypes in order to test the newly designed products in the actual plant. These production runs often proved to be a bottleneck in the development process, but since we did not know the actual procedure for scheduling them, we did not incorporate them in our model.

After collecting specific data for Project X, we turned to the task of generating distributional data for a portfolio of new product projects. The following discussion describes the main methodological and managerial issues we encountered.

Interarrival times. We based our data for job interarrival times on management interpretations of quarterly status reports for the few years preceding our study. Our informants also characterized for us the distribution of project terminations for those projects that were never completed. Typically these terminations occurred toward the end of Phase 3 when engineers determined that no feasible solutions existed for the remaining problems inhibiting scale-up or launch.

Activity times. Using our baseline new product project as a reference and "reality check," we asked for 10th and 90th percentile estimates of the times necessary to complete the activities. That is, we expected that 10% of all occurrences of an activity took less time than our low estimate and 10% took more time than our high estimate. We characterized the prototyping activities according to the experience of the baseline project, since these activities were of relatively short and invariant duration.

In addition to individual task times, we also collected estimates of the total time devoted to activities across all resources. This entry provides some indication of the correlations among activity times of the different resources.

For example, time estimates given for the activity 'Field Trials" show a 10% chance that product engineers contributed more than 700 hours to this activity and also a 10% chance that application engineers contributed more than 700 hours to it. However, the total estimate indicates that the 90th percentile of time spent on this activity by all resources was only 700 hours. Thus if one of the two resources spent a large amount of time on the activity, then the other resource necessarily devoted less time to it: the two task times (Field Trials/Product Engineers and Field Trials/Application Engineers) were negatively correlated. On the other hand, the time an engineer devoted to a particular activity is likely to have been positively correlated with the time her technician spent on it. Such correlations have important implications for project management as well as for process models, but they did not appear to be quantities that our informants could confidently estimate without records of actual project histories.

We encountered similar difficulties in accounting for interdependence among activities as among tasks. One might wonder whether a longer-than-normal Phase 1 is typically followed by a longer-than-normal Phase 2 — that is, if it was difficult to resolve feasibility issues, will it also be difficult to formulate a process plan? Or is the opposite situation more representative — that if the development group spent more time resolving Phase 1 issues, it will be more straightforward for them to formulate process plans? Our model assumed that the two activities are independent. To answer empirically the question of dependence among activities, we would need a joint distribution of all the tasks that compose a product development effort; but the engineering organization we studied did not keep the detailed data necessary for such analysis. It is important to note that in general tasks may be neither perfectly independent nor perfectly correlated.

Likelihood of Iterating. We found that the likelihood of iterating was the most difficult part of our model to specify. To completely characterize it, we would need to describe not only the number of times that activities occurred, but also the order in which they occurred, and we would need probability distributions over both of these differentiating features.

The workshop participants found it difficult to discuss the range of possible configurations that they encountered or to identify characteristic patterns, so we needed to take a simpler route. We asked the engineers to classify 11 recently completed projects according to the complexity of the iteration structure (two projects were simple, five medium, and four complex); we developed profiles for each class; and then we used the weighted average of the resultant profiles to analyze the organization's overall performance.

Implicit in this approach was the assumption of dependence among different iterations in a project. It appeared that in reality some project iterations were independent of one another, and others were negatively correlated; but we could generate only piecemeal indications of what form of data collection might prove most revelatory.

We gathered two forms of data for the iteration structure. For prototyping iterations (those involving the making and testing of materials and products), which were typically repeated many times, we collected "expected total number of iterations per project." This form of data reflected an expectation of strong negative correlation among nested levels of iterations. It also reflected our finding that our informants were typically not well informed on the iteration phenomena outside their domains; we considered it best to ask each informant only for estimates of the numbers of iterations she herself performed. For "major" iterations involving several activities, which occurred in fewer projects and were less numerous when they did occur, we collected data in the form of "probability of iterating," sometimes with a maximum number of times that an activity could be executed. We treated this maximum as a global maximum and treated visits before the maximum number as independent. This raises the following interpretive question: when the maximum is reached, should we assume that further visits to the activity will always succeed, or should we allow the possibility of failure (representing project termination)? We made the first, simpler assumption.

Resource Availability. Finally, we required data that quantify the capacity of the resources. We easily estimated the number of people in each resource pool and the average hours they worked in a week. The time cards enabled us to estimate fractions of time devoted to administrative and support activities by each resource group. An internal study gave us estimates of the average work content of each activity in reformulation projects.

The final kind of data characterizing resource capacity pertains to their functional flexibility. The interchangeability described in the preceding section can be attributed to our choices of resource and activity partitions: either the resources involved have some overlap of function or some set of tasks that we aggregated into a single activity could actually be differentiated. A qualitatively different kind of interchangeability occurs when one resource is over-burdened and turns his work over to someone else. We reflect this phenomenon by partially reallocating the work of over-used resources. Based on conversations with our informants, we estimated the maximum extent to which this reallocation could occur. In some situations, one might suppose that the new person is less well qualified for the task, and that as a result the process time and likelihood of error are increased; but we left this issue for future research.

6. Conclusion

In other papers, we have summarized the operational results from our analysis of this model (reported in Adler et al. 1993a and 1993b); here we focus instead on what we believe we have learned about the features that distinguish the product development process from other types of processes. Exhibit 5 presents a two-dimensional typology of processes. The first dimension contrasts more and less repetitive operations — from this point of view, product development resembles other "professional/creative" activities in that it is much less repetitive than mass production manufacturing operations or routine clerical operations. The second dimension contrasts processes whose key inputs and outputs are material and processes that are essentially concerned with information and knowledge.

		Nature of key inputs and outputs	
		material	information/ knowledge
Degree of Repeti-tiveness	high	CRAFT	CLERICAL
	low	MASS PRODUCTION	PROFESSIONAL/ CREATIVE

Exhibit 5: Four Types of Processes

We can use this taxonomy to classify the characteristics of product development that we identified, to summarize our degree of success in modeling each characteristic, and to sketch future research prospects.

A first group of process characteristics might appear in any type of process, but are not usually critical outside the professional/creative type of process:

- "Servers" in any process might exhibit some load responsiveness, but unless the focus of the study is overtime scheduling, this feature is not likely to be important enough to warrant incorporating into models of processes outside the professional type.
- Iterations are a key feature of some product development environments and some other professional activities, but are not usually a major concern for other types of processes (note, however, the exception of semiconductor fabrication studied by Chen *et al.*, 1988).

• Similarly, flexible pooling might appear in some other types of processes (in the guise of multi-skilling, for example), but it has not been a major point of focus in previous modeling research. In contrast, it appears to be a rather important feature of engineering and other professional environments.
• The possibility of terminating projects in mid-course is an important characteristic of the product development process; any of the other three types of projects might also manifest "yield" problems, but these have not often been the focus of process modeling efforts.
• Interdependencies between phases is a phenomenon that our project skirted because of the added data collection and modeling complexity it would have introduced. But future research might usefully focus on the topic, since difficulties experienced at any one phase often have downstream repercussions. This seems particularly salient in engineering activities, but semiconductor manufacturing exhibits similar interdependencies.

A second group of characteristics does not appear in mass-production manufacturing operations, but may appear elsewhere:
• The uncertainty of task times is likely to be a key issue only in less repetitive processes, whether material-intensive or information- and knowledge-intensive.
• Rapidly changing technologies and product mixes are also characteristic of less repetitive processes. Modeling is useful in these contexts primarily to identify key transition planning issues, but the newness of technologies or products may also lead to insurmountable uncertainty concerning the tasks to be accomplished.
• Forking only appears in a very limited form in material processes (disassembly operations), but is common in any process in which the emergent product can be duplicated, for example using a photocopier.

And finally, a third group of characteristics seems specific to product development and other professional/creative processes:
• Flexible forking — the fact that there is a certain element of management discretion in where to fork — is likely to be a central issue only in knowledge-intensive processes, since the routine nature of clerical information-processing allows managers to determine with much less ambiguity the optimal process flow.
• The incremental release of information to downstream resources would appear to be a potentially important feature of product development (see also Clark and Fujimoto, 1991), but we did not explore it in this project. It could perhaps be modeled effectively with the benefit of more detailed information on the impact of the incomplete information on downstream activities.
• Informal lateral sharing of information is potentially important in knowledge-intensive forking-intensive processes; this issue too we have left for future

research. More detailed precedence diagrams might suffice to capture the essential issues here.

• The interdependencies between activities within a given phase would appear to be an important feature of product development. More detailed data would facilitate modeling this characteristic, but the low degree of repetitiveness makes it difficult to accumulate enough data to form reliable estimates.

It is perhaps not surprising that our model of the product development process was relatively successful for the first group of characteristics, less successful for the second group, and in general unsuccessful for the third. Modeling the product development process will require a more sustained effort by a larger community of researchers.

REFERENCES

Adler, P. S. and B. Borys, "'Tools, Not Weapons': An Inductive Study of Engineers and Formalization," unpublished, 1992.

Adler, P. S., A. Mandelbaum, V. Nguyen, E. Schwerer, "From Project to Process Management in Engineering: An Empirically-based Framework for the Analysis of Product Development," to appear in Management Science, 1993a.

Adler, P. S., A. Mandelbaum, V. Nguyen, E. Schwerer, "From Project to Process Management in Engineering: Strategies for Improving Development Cycle Time," unpublished, 1993b.

Alexander, R., "Scheduling and Resource Allocation Methodologies for Fast Product Development in a Multi-Product Environment," unpublished master's thesis, Sloan School of Management, Massachusetts Institute of Technology, 1990.

Blackburn, J. D., "New Product Development: The New Time Wars," in Time-Base Competition: The Next Battleground in American Manufacturing, J. D. Blackburn, Ed., Business One Irwin, Homewood, 1991.

Chen, H., J. M. Harrison, A. Mandelbaum, A. van Ackere, and L. M. Wein, "Empirical Validation of a Queueing Network Model for Semiconductor Wafer Fabrication," Operations Research, 36 (1988), 202-215.

Clark, K. B., and T. Fujimoto, "Lead Time in Automobile Product Development Explaining the Japanese Advantage," Journal of Engineering and Technology Management, 6 (1989a), 25-58.

——— and ——— , Product Development Performance: Strategy, Organization, and Management in the World Auto Industry, Harvard Business School Press, Boston, 1991.

Dean, B. V. (Ed.), Project Management: Methods and Studies, North-Holland, Amsterdam, 1985.

Hayes, R. H., S. Wheelwright, and K. B. Clark, Dynamic Manufacturing: Creating the Learning Organization, Free Press, New York, 1988.

Schoenberger, R. J., World Class Manufacturing: The Lessons of Simplicity Applied, The Free Press, New York, 1986.

Stalk, G., Jr., and T. M. Hout, Competing Against Time, Free Press, New York, 1990.

Weick, K. E., "Enactment Processes in Organizations," in New Directions in Organizational Behavior, B. M. Staw and G. Salancik, Eds., St. Clair Press, Chicago, 1977, 267-300.

Wheelwright, S. C., "Time to Market in New Product Development," ICL Technical Journal, November (1989), 625-646.

Wheelwright, S. C. and K. B. Clark, Revolutionizing Product Development, Free Press, Free Press, 1992.

Wheelwright, S. C. and K. B. Clark, Creating Project Plans to Focus Product Development, Harvard Business Review, March-April 1992, pp. 70-82.

5

RESOURCE ALLOCATION PROBLEMS ARISING IN PRODUCT DEVELOPMENT

Sriram Dasu
Anderson Graduate School of Management, U.C.L.A.
Los Angeles, CA 90024

1. Introduction

Problems of resource allocation arise when firms are faced with a range of project opportunities but have limited resources to invest in these options. Among the most important characteristics of the various projects are the degree of uncertainty in the resource requirements for the project and the expected pay-off to be obtained on completion. Careful selection of the set of product development projects to pursue in any given time period -- given limits on the resources available for development, including capital, labor and technological know-how -- and control of these resources in the implementation of the development projects are crucial aspects of the task of managing the product design and development process.

A commonly observed characteristic of product development projects in practice is the ambiguity in the statement of the precise requirements of the product at the start of the project. Often there are alternate methods of meeting these requirements and different feasible approaches for implementing each system and its components (e.g. Hubka 82, Hauser and Clausing 88). Selecting the approaches to pursue and determining whether to pursue them in sequence or parallel are among the important decisions which product development managers make.

In this paper we describe a hierarchy of resource allocation problems that product development managers face in practice. We present a descriptive study of one firm which produces a range of electronic products for military and commercial markets. Our focus is on the decisions faced by the managers involved in the development process, and one development project is described in sufficient detail to identify the hierarchy of resource allocation decisions. We abstract the important details of this example as a basis for building the framework.

Our analysis is based on the experiences of a firm that for a variety of reasons could not raise resources to fund each project solely on the basis of its individual merit but had to allocate limited resources across different options. The managers were also particularly concerned about projects that entailed considerable technical uncertainties. The specific project that is the basis of much of the framework was fraught with technical uncertainties.

The product development process can vary quite significantly from industry to industry, and even within an industry there may be substantial differences in the approaches adopted by different firms. These differences may arise from differences in technologies, availability of resources to the firm (skills and number of engineers and scientists, development tools, etc.), organization structure, etc. The framework presented here is certainly not universally applicable. Nevertheless, we believe that there are several firms where the framework will be relevant. In other cases only certain aspects of our analysis may be applicable.

We group the resource allocation decisions into three sets. The first set is concerned with the construction of the portfolio. The objective here is to determine the broad scope and timing of various potential projects, taking into account the resource requirements, pay-offs and risks. Within the broad scope defined at the portfolio level, there often are alternative approaches for developing the product. In the presence of technical uncertainty it may be necessary to explore multiple ways to develop a single product. The second set of decisions is concerned with the order in which the alternatives are to be explored.

Often the product architecture consists of several modules that are interconnected. The developers' degreee of confidence in predicting the capabilities of a module can vary. This uncertainty can be resolved by exploring the modules to a greater degree. Since there are uncertainties about the capabilities of individual modules and their interconnections, there is a need to balance the rate at which different modules are explored. Accordingly the third set of problems, termed the balanced development problems, concerns the pace at which different modules are explored.

1.1. Related Literature. In order to systematize design development, the design process has been studied extensively by researchers belonging to many domains ranging from cognitive sciences (e.g. Simon 70) to engineering (e.g. Suh 70, Hubka 82, Gore 89) to management (Clark and Fujimoto 91, Cusumano and Nobeoka 92, Eppinger et. al. 89, Ulrich et. al. 91). The objective of this literature is to aid development of good designs, taking into account a host of factors such as translation of customer requirements into technical requirements (e.g. Hauser and Clausing 88), generation and selection of functional specifications and physical elements that meet the technical requirements (e.g. Suh 70, Hubka 82, Gero 89, Sriram et. al. 89), manufacturability (e.g. Nevins and Whitney 89, and Susman 92), and design development time (e.g. Eppinger et. al. 89, Ulrich et. al. 91, Clark and Fujimoto 91). The focus throughout this body of literature is on the myriad of issues that arise in the development of a single product.

In the problem we encountered, the firm simultaneously had to upgrade several products and launch new ones. Therefore we are also concerned with the problem of selecting, organizing and managing a dynamic portfolio of design projects.

Wheelwright and Clark (92) find that the problem of managing a portfolio of product development projects is encountered frequently. The authors observe that many companies have no formal process for choosing and managing the development projects and argue that management should make sure to link the mix of product development projects the firm undertakes, and the sequence in which they are executed, directly to the overall strategy of the firm.

While Wheelwright and Clark provide a broad framework for constructing a product development portfolio, they are not directly concerned with problems of resource allocation. The problem of constructing research and product development portfolios has also been studied by researchers in many fields including operations research and economics. Deshmukh and Chikte (80) classify this literature into two groups -- micro and macro. The "macro" literature is concerned with the effects of firm size, competition, and market structure on the nature of investment in research and development and the rate of innovative activity (e.g. Kamien and Schwartz 82, and Reignanum 89). The "micro" literature focuses on problems within the firm, and is concerned with the problem of either allocating resources across competing projects (Baker 74, Fox et. al. 84, Ali et.al. 93) or allocating resources to various phases of a selected project (Nelson 59,61, Abernathy and Rosenbloom 68, Deshmukh and Chikte 77, Bard 85, and Posner and Zuckerman 90). This paper belongs to the "micro" literature.

In sum, the existing literature on product development can be grouped into two types. One type is concerned with the product development process and the other is concerned with the construction of product portfolios. The first type focuses on the detailed structure of designing a product, while the second type does not take the detailed structure of the design process into account.

This paper attempts to links the detailed structure of the design process to the larger or higher level problem of product portfolio determination. We consider a spectrum of resource allocation and sequencing decisions that arise in product development, ranging from the problem of allocating resources across competing projects to sequencing and scheduling decisions that arise in different phases of the product development process. In addition to the range of issues considered and the resulting interconnections that are revealed, the structure of the individual problems also differs from those studied by other researchers. For instance we view the problem of allocating resources to competing projects to be a dynamic problem where the firm must determine not only which options to pursue but also the sequence in which to pursue them.

To develop the framework, in the next section we describe the case of a firm in the electronic industry. We describe its existing portfolio of products and the procedures it employs in identifying new opportunities for product development. To illustrate the hierarchy of resource allocation decisions encountered, and the methods used to cope with uncertainty in practice, we describe one project in greater detail. In the third section, we abstract the details from this descriptive analysis which are important in modeling the resource allocation decisions. In the fourth section we develop mathematical models. For reasons of confidentiality we have altered the name of the company.

2. Illustrative Example

EXT Electromagnetic Instruments Division (EID) was an independently managed subsidiary of a large conglomerate. Founded in 1945, its first product was a vacuum tube based electromagnetic interface receiver developed for military application. The firm grew rapidly by exploiting advances in the theory of electromagnetism and the design of electromagnetic circuitry to produce an increasingly broad portfolio of instruments for generating, receiving, measuring, modulating and transmitting electromagnetic signals. Included in the portfolio were:

(a) Frequency synthesizers used for scrambling electromagnetic transmissions. Typical applications include use in systems for protecting communication transmissions, tracking satellites, and detecting aircraft and missiles before they appear on the horizon.

(b) Spectrum analyzers used to receive and analyze microwave and electromagnetic transmissions in order to determine precisely the power and frequency of each of the transmitted components of a complex signal. These instruments are used in military surveillance equipment, and for testing microwave devices at the point of fabrication.

(c) "Auto Control Instruments" used for calibrating circuits in auto-pilot devices for steering aircraft and missiles in precisely specified directions.
(d) Broad band power amplifiers capable of amplifying very low-powered signals over a very broad frequency range by a large and precise factor. These are used for testing ultrasound medical imaging systems, anti-submarine warfare systems, and for ultrasound cleaning.
(e) Electromagnetic interference probes, receivers and antennas used for measuring the interference emitted by electrical and electronic components in devices as varied as navigational guidance systems in submarines, radar communication equipment, telemetry and avionics systems in ships and aircraft, and commercial products such as televisions, VCRs, TV games (e.g. Nintendo), Xerox machines and office automation equipment.
(f) Electromagnetic susceptibility test systems to test electromagnetic compatibility among different electrical and electronic systems that must operate together in one room or vehicle. Primary buyers are in the aerospace industry.
(g) Noise sources and noise measuring instruments used in testing electronic components. They are purchased by manufacturers of TVs, VCRs, radar receivers, and cordless telephones, for use on the production shop floor, making this is a high volume product line for EXT.

Families (b) and (c) were thought to have very little potential and EXT planned to stop producing them. Families (a), (e), (f) and (g) were targeted as strategic growth areas.

The firm offered a broad range of products employing technologies that spanned several decades. Several of the products were used as subsystems by the defense industry. In many cases the primary system (in which the EXT subsystem was embedded) was developed several decades earlier and was not updated since. EXT was obliged to meet the replacement demand for these subsystems. EXT in turn had not updated its design of these old subsystems and therefore used parts that were obsolete. These parts had to be special ordered from semi-conductor manufacturers, resulting in long procurement lead times and high costs.

Competition had intensified in many of the firm's key markets as new competitors entered and offered competitive products at lower prices. To enhance its competitive position EXT was compelled to upgrade the design of its older products. Product redesign was expected to benefit the firm in two ways. First, replacement of older obsolete parts by new parts would reduce procurement lead times and costs. Second, through redesign, several of the products could be consolidated into a smaller set. Due to advances in technology it was possible to design a product that has the same capabilities as several older products. Thus the set of products being produced could be reduced without withdrawing from any of the market segments. The reduction in the number of products would generate considerable efficiencies in procurement, manufacturing and marketing.

EXT decided to redesign several of its products. However, due to declines in the defense budget and the increase in competition, EXT realized a significant decline in its sales revenues. The financial and engineering resources available to the firm were highly constrained. Hence it was important for EXT to select and sequence the product development projects carefully .

To better understand the nature of the technological uncertainties,and its impact on the resource allocation decisions, we provide a brief description of a recently completed project.

The materials available for the study were official project documents, informal memoranda and some interviews. The written coverage of events, maintained in a project log by the engineering manager, was sometimes sketchy, and the time available for interviewing to fill the gaps was somewhat limited. It is fair to say that for the major events in the history of the project the available records were examined, and for others, a sample of the available records were examined to clarify the account provided by the engineering manager.

2.1. The Cobra Project. The project was associated with the synthesizer family of products. The synthesizers receive and transmit radio signals. The frequency at which these transmissions are made changes constantly in a predefined manner. The goal of the project was to develop a new ultrafast synthesizer capable of outperforming existing counterparts in the market. The product was to be designed with the existing customer base in mind, and its creation would have the effect of making some of the existing synthesizers obsolete. A project of this nature corresponds to the development of a new "core" product for an existing family of products.

The initial concept originated from the marketing group, spurred in part by the encroachment of competitors who had introduced new synthesizers capable of many of the same features, and in part by the difficulties encountered in trying to customize the existing family of products for new customer applications.

Phase 1 : Defining the Project Scope. When this project was initiated, EXT offered three series of synthesizers consisting of eleven basic products. These products often had to be further customized. Thier key attributes were the operating frequency range, switching speeds between different frequencies, signal to noise ratio, phase modulation capability, and real time correction of output power. There were four competitors in this industry and EXT products dominated the market in terms of technical capabilities.

The objectives of the redesign effort were to :
(a) enhance the switching speed and phase modulation capabilities, and
(b) consolidate the product range so as to reduce the eleven products into no more than four products that could be easily customized using software rather than hardware modifications.

High-end EXT synthesizers could switch frequencies every 20 microseconds (μsecs), while the competitors' products could switch frequencies every 40 to 60 μsecs. Thus EXT products were more than twice as fast as those of their competitors. However, it was believed that within a year one of the larger competitors would bring to the market a synthesizer with switching speeds of 10 μsecs. EXT therefore decided that its next generation of synthesizers should be capable of switching frequencies at least every 5 μsecs if not faster. They expected that competitors would require at least three years to reach a 5 μsecs speed and six years to reach 1 μsecs speed. One of the major issues in defining the scope of the project was whether to target switching speeds of 5 μsecs or 1 μsec.

The second major variable that defined the scope of the project was the operating frequency range. The objective of the redesign was to reduce the number of offerings by designing broad band synthesizers. The new products would also have software capability to store predefined frequency ranges, thereby accelerating the performance of the synthesizer and facilitating customization. The complete operating range was 10 Khz to 2.56 GHz. One of the issues that had to be resolved was whether to divide this region into one, two or three (overlapping) ranges.

As the switching speeds and the operating frequency ranges increased the noise levels also increased. Hence to enhance the capabilities several technical problems had to be solved. In particular, if a switching speed of 1 μsecs was targeted, then several significant changes in circuitry were necessary. The engineering manager initially estimated that if nine design engineers allocated a significant portion of their time to this project, then a synthesizer with a frequency range of 10 Khz to 2.56 GHz and a switching speed of 1 μsec could be developed in 24 to 30 months. A product with the same frequency range but a switching speed of 5 μsecs could be developed by the same team in 15 to 18 months. He also believed that the development of the slower product would provide only some incremental knowledge toward the development of the faster synthesizer. This is because the slower system could be developed by refining existing technologies, while the transition from 5 μsecs to 1 μsec would require at least minor modification in most subsystems and significant redesign of three critical circuits.

Management opted to devote resources to the development of a synthesizer with a frequency range of 10 Khz to 2.56 Ghz and a switching speed of 1 μsec.

Phase 2: Product Architecture and Subsystem Specification. The first task in the design process was to transform the loosely specified requirements into an explicit set of design specifications. In generating these specifications, the engineering manager and a team of experienced designers used knowledge gained in the previous projects to guide decision making and to determine the attributes of the subsystems.

Within the space of performance requirements, many feasible design alternatives remained. Preliminary configuration diagrams which illustrated the major subsystems of the product and the logical relationship between them were developed for a number of these alternatives by the engineering group. Some of these alternatives were more conservative than others in terms of the technological know-how required. Some would involve refinements of design using existing technologies, and although there was a high degree of confidence in the firm's ability to implement these designs, there was less confidence that these approaches would yield additional improvements in the future. Other alternatives would require experimentation with new techniques with which the firm had limited experience. Solutions of these technical problems, however, could lead to products whose performance could be enhanced and expanded over time.

For example, one alternative was to develop a direct digital approach to phase modulation. This approach would entail extending the state-of-the-art in the design of one module (phase-lock loops) and would enable exceptionally pure output with very low noise (phase noise, white noise, very low harmonics and sub-harmonics) .

For each system configuration, it was necessary to develop more detailed specification consisting of upper and lower bounds on performance requirements for all major subsystems. These requirements were specified in a manner which ensured that the major subsystems were consistent with each other and with the overall objective of the product.

An important task during this phase of the design process was to develop analytic models and perform preliminary engineering analyses to evaluate the feasibility of the different alternatives and to assess the difficulties which were likely to be encountered in attaining the desired performance requirements. These results were used in turn to estimate the resources required for implementing the subsystems. Because of expenses associated with each of these analyses, several potential alternatives were not considered.

Among the important issues with which the engineering manager was faced at this phase of the process were (a) the amount of resources to allocate to evaluating each of the selected subsystems and (b) the order in which alternative methods of implementing these subsystems should be evaluated, including whether to pursue any of these alternatives in parallel.

Phase 3: Subsystem Evaluation and Development. In this phase, the subsystems were further explored and developed for each of the alternative configurations,. Subsystems which were felt to be easily attainable were not subjected to the same level of analysis in phase 3. Instead attention was focused on those aspects of the design that were felt to be critical to the overall performance of the system and where there was technological uncertainty. For these subsystems preliminary models, referred to as bread-board models, were created to evaluate the feasibility of implementing certain technically challenging alternatives.

The level of detail in developing the specifications at this stage was a function of the variations in the degree of uncertainty about being able to develop the components, and the variations in "system criticality" of the components. By this we mean the extent to which the required performance level of one component affects other components in the system. If the performance of a given component affects the required performance level of a great many other components, then care was taken in prematurely issuing detailed requirements of these "system critical" components, even if the developmental difficulties associated with the critical components were only moderately high.

The decision as to the appropriate level of detail for specifying components at this or any other phase of design required balancing two factors: (a) the risk of having to revise entire systems when detailed predictions about the component capabilities or characteristics turn out to be wrong, and (b) the risk of having to revise the detailed specifications due to changes in the subsystem requirements. Here the primary link between the different subsystems was due to the extremely tight restrictions on the noise levels and the very high switching speeds.

To resolve the uncertainty at this stage, the manager again chose to pursue additional experiments and paper analyses. The important class of decisions which he faced concerned the order in and the extent to which these alternatives were to be pursued. These decisions were influenced by the availability of engineering resources and whether or not the resources could be split among several subsystems.

Phase 4: Detailed Final Design. The task in the fourth phase was to specify the characteristics of individual devices and circuits used in producing the components which comprised the different subsystems of the products, and to develop detailed schematic diagrams illustrating the logical and physical relationship between these various devices. At this level of detail individual electronic circuits were represented in the drawings.

Once the specifications had been established, a schedule for the remaining stages of the project was developed showing:

- expected completion time of a working prototype (referred to as the brass-board model)
- release of manufacturing specifications.

In developing the detailed design for each device, there were again several possible alternate implementations which would satisfy the specifications. Since the costs and capabilities were known with greater certainty, it was now possible for the marketing department to more accurately assess the market potential. In the Cobra project there was an interesting link between the manufacturing costs and the sales volume. With higher levels of sales it was economical to increase the level of integration in the circuits.

The Project Outcome. The Cobra project resulted in a very advanced synthesizer that contained several patented circuits. It achieved all its technical goals with an operating frequency range from 10 Khz to 2.56 GHz, less than 1 microsecond switching speed, exceptional spectral purity, modular construction for easy customization and high reliability, and a menu driven user friendly software system. However, the project required close to 48 months and exceeded the development budget by 150%. It absorbed much of the firm's engineering resources, in particular the attention of senior engineers. As a result several other development projects were inadequately funded and staffed, and were significantly behind schedule. As a consequence EXT's competitive capabilities were seriously compromised. Eighteen months after the completion of the Cobra project, the parent corporation disbanded this sub-division!

Several causes contributed to the long development time and high cost of the Cobra project. There was high turnover in the engineering staff and the development team size varied considerably over its lifetime. The number of engineers working on the project ranged from four to twelve. Some parts of the development, including the software development, were subcontracted out. The interface between the hardware and software was not effectively managed, resulting in many revisions in both the hardware and software designs. Several hardware modules had to be completely revised several times. During the course of the development, the design team identified new technical possibilities and consistently biased all goals toward the high-end.

In effect management was inconsistent in the priorities it assigned to different projects, and this affected the funding levels of different projects. Management lacked an aggregate project plan (Wheelwright and Clark 92) to guide them in assigning priorities. Engineers may have wasted time and resources exploring alternatives in a disorganized manner. Engineers also appear to have focused excessively on extending the state-of-the-art perhaps because they lacked clear objectives in terms of product capabilities and completion deadlines. The lack of rigid guidelines can also be attributed to the absence of an aggregate project plan.

3. Abstraction

From the preceding description of the design process, we observe that the space of performance requirements in which the engineers work is combinatorially very rich. This richness forces the designers to be selective in their search process-- it would be impossible to examine and evaluate all of the possibly infinite number of potential solutions.

The essential features of the strategy outlined above for dealing with this complexity can be summarized as follows. At the initiation of any given development project the engineering and design managers, in collaboration with the marketing manager, identify a space of possible performance capabilities and estimate the development and production costs for these alternatives. They also develop estimates of the "pay-off" associated with each region in the performance space and the time window in which the products must reach the market to maximize payoffs. At the outset these estimates may be coarse. Within the chosen region, an initial set of performance specifications is then chosen as the system objectives. The system is decomposed into a number of subsystems, and detailed specifications are developed for the technically more difficult subsystems. The number of ways to implement solutions to these subproblems may be enormous and only a few can be explored. To begin with, several broad technical alternatives for solving these problems are considered. Through detailed analysis and model building the number of alternatives is narrowed as the alternatives are explored and knowledge of the technical problems is gained. Several alternatives may be explored when there is considerable uncertainty about the potential success of any single approach. Gradually, a choice of one technical alternative and one set of component specifications is made. Convergence on the final solution is preceded by several iterations among the different phases for modifying the detailed designs and the subsystem specifications.

As has been observed by many researchers (Simon 70, Pahl and Beitz 88), the process of attacking the design problem therefore involves decomposing the tasks into a number of smaller problems, and synthesizing the solution for the larger problem from the solution to the smaller problems.

The process of decomposing the design problem and searching for a solution generates knowledge (new alternatives) and consumes development time and engineering resources. Once the problem is decomposed into subproblems it may be possible to explore the subproblems in parallel. However, while parallel exploration may reduce the development time, it may increase the resources required. Also, parallel exploration will be constrained by dependence between the subproblems. Depending on the strength of the linkages between the different subproblems, it may not be possible to make firm commitments to subproblem solutions until all the related subsystems have been successfully solved. Thus it may be wasteful to explore a subsystem in greater detail until the related systems have been studied to an appropriate degree. Even if the subproblems have minimal dependency, it still may be more appropriate to first explore subproblems that a priori are expected to be technically more problematic. The order in which the problems are explored will affect the resources that are required, the time to develop the product, and possibly the quality of the solutions that are obtained. Thus the firm needs to balance judiciously the pace at and the order in which subproblems are explored.

Cobra project also points to the need to balance carefully the set of projects that the firm is exploring. This was a high-risk project that also had the potential for very high pay-off. But, given the financial status of the firm, the strategy was excessively risky. The risk could have been mitigated by either reducing the scope of the project or by pursuing additional development projects which had lower risks.

In the next section we present models that formalize the trade-offs confronted by the managers in EXT.

4. Mathematical Models

Based on the preceding discussion the set of resource allocation and sequencing decisions encountered in EXT can be cast into a hierarchical framework. At the highest level of the hierarchy we have the product development portfolio problem (aggregate project planning). Here broad alternatives within each family are considered for funding and scheduling. At the next level in the hierarchy, decisions have to be made about the number of alternative configurations to be considered in each project and whether to consider these alternatives in sequence or in parallel. The third set of problems arises while exploring the alternatives. Here we have to ensure that development occurs in a balanced manner (the balanced development problem). In the balanced development problem trade-offs have to be made between (a) the risk of having to revise the entire system (or several subsystems) when detailed prediction about the component capabilities or characteristics turn out to be wrong, and (b) the risk of having to revise the detailed specifications due to changes in the subsystem requirements.

Aggregate Project Planning: Since this is the highest level in the hierarchy, the alternative projects considered here will be the broad set of alternatives for each product family. For instance in the synthesizer family there were at least four broad alternatives, depending on whether the switching speed was targeted at 5 or 1 microsecond, and whether the operating frequency range of 10 Khz to 2.56 GHz was partitioned into three regions or a full range synthesizer was developed. As discussed in the previous section, within these alternatives many possibilities still remain. However, for the purpose of the aggregate plan it will suffice to consider just these broad alternatives.

Associated with each alternative will be resource requirements and potential revenues that will accrue upon project completion. The revenues will clearly depend upon the project completion time. We can also expect the resource requirements to depend on the sequence in which the projects are executed. Recall that in the Cobra project the development of a 5 μsec synthesizer had some commonality with the 1 μsec synthesizer. If the development of a 5 μsec synthesizer preceded the 1 μsec product, then some reduction in the resources requirement and the uncertainty for developing the 1 μsec product could be expected.

The purpose of the aggregate plan is to determine for each family which of these alternatives to pursue and how to schedule them, taking into account the resources that are available. Depending on the circumstances, several criteria may be employed in selecting from the set of feasible schedules. We present below one possibility, where the objective is to maximize the expected discounted revenues and the constraints include bounds on the risks faced by the firm. In the formulation given below considerations of risks are incorporated by requiring that some percentile of the revenue distribution never drops below a certain limit. Several other alternatives can be considered including a bound on the squared coefficient of variation of the revenues in each period.

Maximize

$$\sum_{t=1}^{T}\sum_{f} \rho^{t-1} E(P_f(t)) \qquad \text{(Expected Revenues)}$$

Subject To:

Completion time of project :

(1) $c_{f,j} \;\;= s_{f,j} + d_{f,j}(k_f(s_{f,j}))$

Revenues from a completed project :

(2) $P_{f,j}(t) = \begin{cases} 0 & \text{if } t \leq c_{f,j} \\ p_{f,j}(t) & \text{otherwise} \end{cases}$

The state of development :

(3) $k_f(t) \;= Max\ \{j:\ c_{f,j} \leq t\}$

Revenue only from the most advanced product :

(4) $P_f(t) \;= Max\ \{P_{f,j}(t)\}$

Resource constraint:

(5) $\sum_{t} r_{f,j}(t - s_{f,j}) \leq R$

Lower bound on revenue :

(6) $\sum_{f} E(P_f(t)) - a\sigma(\sum_{f} P_f(t)) \geq L$

Where

a	: a constant
$c_{f,,j}$	: completion time of version j in family f
$d_{f,,j}(k)$	: development time of version j in family f given that version k has been developed
E(.)	: Expected value
$k_{f,,}(t)$	: index of the most advanced version of family f developed by period t
L	: a constant
$p_{f,,j}(t)$	: revenue from version j in family f in period t
$r_{f,,j}(t)$	: resources required by version j in family f, t periods after its start

R : resource available
$s_{f,,j}$: starting time of version j in family f
T : planning horizon
ρ : discount rate
$\sigma(.)$: standard deviation.

Equation 1 determines the completion time of a project. Equation 2 ensures that revenues from a version are available only after that version has been developed. The above formulation is based on the assumption that the different versions within a family can be ordered hierarchically. Versions with higher index numbers are considered to be technically superior. Equation 3 identifies the highest version that has been developed by time t. This index acts as a surrogate for the expertise developed by the firm and thus has an effect on the time it will take to develop other versions. Equation 4 is based on the assumption that at any time the firm only has one version in the market, and the version that is in the market is the one that generates the highest revenues. Resource constraint (5) ensures that in any time period the resources required by the active projects do not exceed the available resources. Finally constraint (6) ensures that the likelihood of the revenues dropping below a threshold L does not exceed some probability. This probability depends on the value of the constant a. The decision variables for this problem are the starting time for each project.

Note that this problem is similar to a multi-resource stochastic scheduling problem (Calier and Rinnooy Kan 82, Igelmund and Radermacher 83) with a side constraint on the revenues (constraint 6). Solution procedures for this problem are discussed in Dasu (93).

Search Problems (Parallel vs. Sequential Approaches) At this level of hierarchy it is assumed that the broad system parameters are defined and decisions have to be made about the alternative system configurations to be explored and the values of secondary parameters. It is also assumed that the resource availability and the target completion times are given.

In the Cobra project the decision to build a broad band synthesizer with a switching speed of 1 μsec was made, and the decision variables to be explored included alternative approaches to designing the filtering networks, the phase-lock loop circuits, the amplifiers etc. One of the broad alternatives was to develop a direct digital design of the synthesizer. This decision would have an effect on several design parameters such as (a) the speed and accuracy with which the output power of the signal is maintained while frequencies are being switched, (b) noise levels (phase noise, spurious, etc.), and (c) ability to modulate the signal while the frequencies are being altered.

At this stage in the development process the primary concern is to establish the viability of an option by determining basic capabilities rather than developing a detailed design. Preliminary engineering analysis and experiments are carried out to establish technical feasibility/capability.

The problem of exploring alternative design strategies has been formulated and studied by several researchers including Nelson(62), Abernathy and Rosenbloom (69), Bard (85), Ali et al. (93). This body of literature evaluates the options at a very high level of abstraction. Although the development process of each option may be viewed as consisting of several stages, no specific reference is made to the architecture of the problem being explored.

It is well known (e.g. Kusiak and Szczerbicki (90), Kott et. al. (90), Eastman) that the alternatives available to a product designer can be expressed in the form of And/Or graphs. For instance to develop a product sucessfully we may require one of the approaches A_i i=1to n to succeed. In order for A_i to succeed we may require modules $A_{i,j}$, j =1 , J_i to be feasible. For each module $A_{i,j}$ there may be several alternative approaches: $A^k_{i,j}$ k = 1,..κ_{ij} and only a subset of these may have to be feasible for A_{ij} to be feasible (Figure 1).

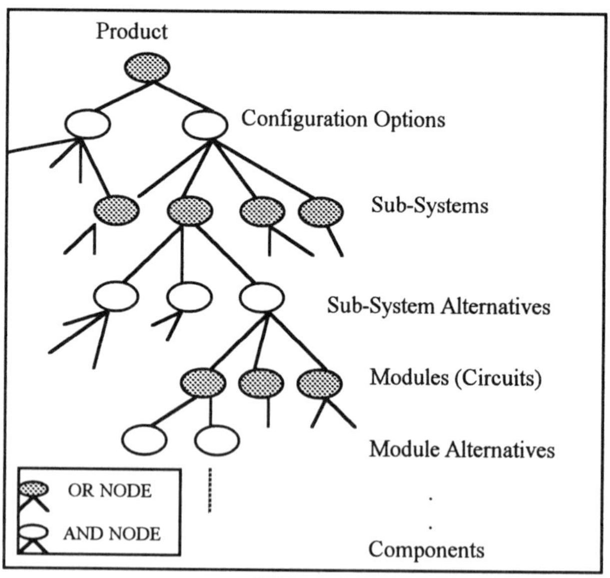

Figure 1

Associated with each alternative are estimates of the likelihood of success and the time required to explore the option. Given these estimates the optimization problem is to determine a sequence for examining the options that maximizes the likelihood of developing a feasible product in the given time constraints.

The problem can be formulated as a dynamic program and branch and bound techniques can be employed to determine a schedule (e.g. Ben-Dov 81, Henig and Simchi-Levi 90, Dasu 93).

As the development process progresses we can expect the technical uncertainties (variance in the estimates of the performance capabilities) associated with the product to decrease, although not necessarily monotonically (Figure 2). The optimal search strategy on the And/Or graphs has the desirable property that it decreases the uncertainty at the fastest rate (Dasu 93).

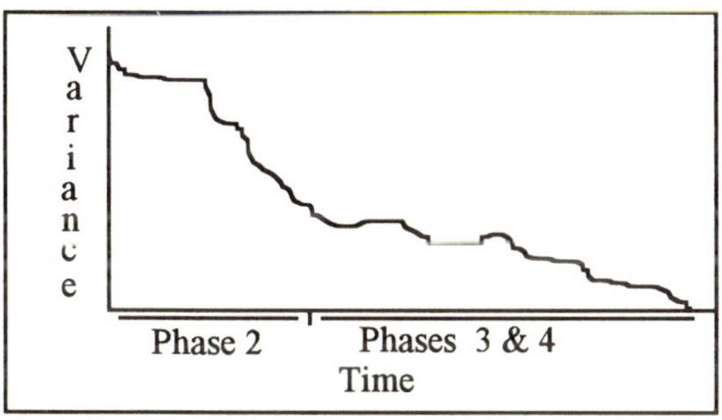

Figure 2

Balanced Development Problem: In the presence of technical uncertainty the initial objective will be to establish preliminary capabilities, and each subsystem may be explored largely in isolation. After the engineers develop sufficient confidence in an alternative, the design process is advanced to the next stage, where the different subsystem capabilities are defined with greater precision and attempts are made to integrate the subsystems. At this stage it is necessary to ensure that the different modules function together and to increase the level of detail with which circuits/modules are specified. Some of the issues that arise during the detail design phases (3&4) are captured by the balanced development problem.

Although by this stage broad capabilities of the alternative options are known, considerable uncertainty about the exact performance of the product remains. In the Cobra project, for instance, the noise levels and output power levels were critical variables. These variables were affected by several factors including the layout of the circuit and the components used. There were three critical subsystems that had the largest impact on performance. In some subsystems performance could be accurately predicted on the basis of the functional design, while in other subsystems it was necessary to build bread boards and test them.

Phases 3 and 4 were the longest phases in the Cobra project and there was considerable concern that the process may have been inefficiently managed. As stated earlier there were many explanations for the over-runs. One of the issues was whether time was wasted by unnecessarily developing detailed designs (bread boards) that were subsequently changed when subsystem requirements were modified. In other words were detailed designs developed prematurely for some subsystems ?

The trade-offs involved here can be conceptualized as follows. Assume for ease of exposition that the design process consists of only two stages. Let us call these high level and low level. Assume that there are n modules. At the high level the interface between the modules is determined and specifications are developed for each module. The specifications are based on judgments of what is attainable. However, due to technical uncertainties there is no assurance that the desired specifications can be attained at the detailed level, or, even if the specifications are attained by each module, that the overall system will work. It is only when the detailed design is developed that feasibility can be ascertained.

Assume that for any specification there is a probability (α_i) that module i is feasible and, given that all modules are feasible, let β be the probability that the system is feasible. Figure 3 illustrates the cycles. Since high level and low level design development activities are time consuming, we need to determine if the detailed design should be developed for all modules concurrently or if detailed design for the modules with low probability of success should be developed first before resources are invested in the more certain modules.

Developing modules in parallel may decrease the product development time but will increase the resources consumed by the project. Hence there is a trade-off between the total resources required and the completion time. Since many projects may be in progress at any given time, resources can be diverted between projects. Hence it is also important to consider the total resource requirement within a project and not just the time to complete the development.

The balanced development problem can conceptually be generalized to several design stages together with the possibility of integrating designs at different stages of development. In general we may even have the option of integrating a sub-set of the modules before testing the entire system. Each integration stage can be viewed as the development of a prototype. Typically there is a sharp increase in knowledge after the completion of a prototype. In the balanced development model learning can be incorporated by reducing the probability of failure and/or decreasing the time to develop each phase.

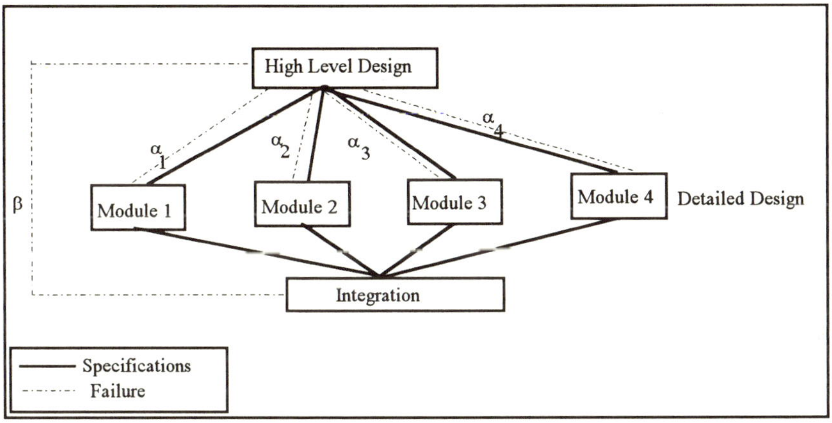

Figure 3

The Cobra project entailed combining analog and digital circuits; hence there were no tools to test designs at different stages of development. For purely digital circuits there are many simulation tools that can test functional designs, logical designs, and help with layouts. It is also possible to verify through computer simulations, even at intermediate stages of development, whether or not the subsystems are compatible (Gajski et. al. 92). Similar problems of linking development phases and integrating subsystems have also been observed in the design of mechanical systems (e.g. Eppinger et. al 89, Bond and Ricci 89).

5. Conclusions

The design development problem can be viewed as a process of converging from a broad, vaguely defined concept to a set of detail specifications (e.g. Hubka 82). Most development projects involve several design iterations. The number of iterations depends on the level of technical advancement required by a project. The process of searching for solutions, and the associated iterations contributes to the risk and the potential pay-offs from a project. In this paper, we have presented the experiences of one firm. Based on this case we propose a hierarchy of problems that must be solved to balance the risks across projects and minimize the resource requirements within a project.

REFERENCES

Abernathy, W.J. and R. S. Rosenbloom (1968), "Parallel and sequential R & D strategies: Application of a simple model," I.E.E.E. Trans. Eng. Man, 15, 2-10.

Abernathy, W.J. and R.S. Rosenbloom (1969), "Parallel strategies in development projects," Man. Sci., 15, B486-B505.

Ali, A., M.U. Kalwani, D. Kovenock (1993), "Selecting product development projects: Pioneering versus incremental innovation strategies," Man. Sci., 39, 255-274.

Baker, N.R. (1974), "R&D project selection models: An assesment," I.E.E.E. Transactions on Engineering Management, 165-171.

Bard, J.F. (1985), "Parallel funding of R&D tasks with probabilistic outcomes," Man. Sci., 31, 814 - 828.

Ben-Dov, Y. (1981), "A branch bound algorithm for minimizing the expected cost of testing coherent systems," European Journal of Operational Research, 7, 284-289.

Bond, A.H. and R.J. Ricci (1989), "Cooperation in aircraft design," in Computer-Aided Cooperative Product Development, Edited by Sriram, D., R. Logcher, and S. Fakuda, Springer-Verlag, New York.

Carlier, J. and A.H.G. Rinnooy Kan (1982), " Scheduling subject to non-renewable resource constraints," Operations Res. Lett. 1, 52-55.

Clark, K.B. and T. Fujimoto (1991), Product Development Performance, Harvard Business School Press, Boston, Massachusetts.

Cusumano, M.A. and K. Nobeoka (1992), "Strategy, structure and performance in product development: Observations from the auto industry," Research Policy, 265-293.

Dasu, S. (1993), "Resource allocation and sequencing decisions arising in product development: A case study in the electronic industry," UCLA, Los Angeles.

Deshmukh, S.D. and S.D. Chitke (1980), "A unified approach for modelling and analyzing new product R&D decisions," TIMS Studies in Management Sciences, 15, 163-182.

Eppinger, S.D., D.E. Whitney, R.P. Smith and D.A. Gebala (1989), "Organizing the tasks in complex design projects," in Computer-Aided Cooperative Product Development, Edited by Sriram, D., R. Logcher, and S. Fakuda, Springer-Verlag, New York.

Fox, G.E., N.R. Baker and J.L. Bryant (1984), "Economic models for R&D selection in the presence of project interaction," Man. Sci., 30, 890-902.

Gero, J.S. (1989), Artificial Intelligence in Design, Springer-Verlag, New York.

Gajski, D.D., N.D. Dutt, A. C-H. Wu, and S. Y-L. Lin (1992), High Level Synthesis, Introduction to Chip and System Design, Kluwer Academic Publishers, Boston.

Hauser, J.R. and D. Clausing (1988), "The house of quality," Harvard Business Review. May-June, 63-73.

Henig, M.I. and D. Simchi-Levi (1990), "Scheduling tasks with failure probabilities to minimize expected costs," Naval Research Logistics Quarterly, 37, 99-109.

Hubka, V. (1982), Principles of Engineering Design , Butterworth Scientific, London.

Igelmund, G. and F.J. Radermacher (1983), "Preselective strategies for the optimization of stochastic project networks under resource constraints," Networks, 13, 1-28.

Kamien, M.I. and N.L. Schwartz (1982), Market Structures and Innovation, Cambridge University Press, Cambridge, Massachusetts.

Kott, A., G. Agin and D. Fawcett (1990), "Configuration tree solver: A technology for automated design and configuration," Advances in Design Automation, 317-322.

Kusiak, A. and E. Szczerbicki (1990), "A formal approach to design specifications," Advances in Design Automation, 311-316.

Nelson, R.R. (1962), "Uncertainty, learning and economics of parallel R and D efforts," Review of Economic Studies and Statistics.

Nevins, J.L. and D.E. Whitney (1989), Concurrent Design of Products and Processes, McGraw-Hill, New York.

Pahl, G. and W. Beitz (1988), Engineering Design, Springer-Verlag, New York.

Reinganum, J.F. (1989), "The timing of innovation," in Handbook of Industrial Organizations, Vol 1, pp 849-908, Edited by R. Schmalensee and R.D. Willig, North-Holland, New York.

Simon, H.A. (1970), The Science of the Artificial, M.I.T. Press, Cambridge, Massachusetts.

Suh, N.P. (1978), The Principles of Design, Oxford University Press, New York.

Susman,G.I. (1992), Design For Manufacturing, Oxford University Press, New York.

Ulrich, K . and D. Sartorius, S. Pearson, M. Jakiela (1991), "Including the value of time in design-for-manufacturing decision making," Working Paper WP#3243-91-MSA, Sloan School of Management, M.I.T. Cambridge, Massachusetts.

Wheelwright and Clark (1992) Harvard Business Review.

6

DESIGNING PRODUCTS AND PROCESSES FOR POSTPONEMENT

Hau L. Lee
Dept. of IE-EM, Stanford University
Stanford, CA 94305.

Corey Billington
Hewlett-Packard Company
3000 Hanover Street, Palo Alto, CA 94304.

Abstract: Companies that compete in a global market have to manufacture and configure products to meet the diverse needs and demands of customers. Such diverse needs can result from geographical differences, government regulations, or simply differences in customer preferences and tastes. Uncertainty in the market place makes it extremely difficult to forecast product demands accurately. This results in high inventory costs and low product availability. Postponement is a strategy whereby the final configuration of a product is delayed as much as possible, usually until a customer order is received. Postponement ranges from simple packaging and labeling, bulk packing or special customization, to more complex forms such as localization, assembly and test, and final product integration. Effective implementation of postponement requires careful redesign of the products and the processes involved. In this paper, we describe a conceptual framework for design for postponement, and the appropriate cost drivers that can be used to assess postponement strategies for a company.

1. Introduction

In recent years, there has been increased emphasis on the incorporation of manufacturing considerations in the design phase of a product. It has been found that the product's manufacturing cost is largely determined by the design of the product. Consequently, it is important to include considerations of the costs, efficiency and quality aspects of manufacturing in the design phase. Concepts such as "design for manufacturability," have emerged both in research and in practice (see Whitney, 1988, Dean and Susman, 1989, and Taguchi and Clausing, 1990). We note, however, that manufacturing is one step in the order fulfillment cycle of the product. We contend that product and process designers should look beyond manufacturability to order fulfillment in their considerations.

Today's market place is marked by diverse customer tastes and preferences, globalization of the market, and rapid technological advances. These factors, as described below, led to a proliferation of product options and models which, in turn, present major challenges to operations managers. While there have been increased pressures for marketing departments to control the proliferation of product variety (see Fisher, 1992, for an excellent review of such a problem), companies are beginning to discover that great opportunities can be found in the design of products that provide much greater efficiency and flexibility in the logistics and distribution aspects of the order fulfillment cycle. Lee (1991) coined the term "design for supply chain management" to describe such design concepts. In this paper, we describe and discuss one such concept, known as "design for postponement."

2. The Curse of Product Variety

In a global market, due to the different local requirements in taste, language, environment and government regulations, multiple versions of a single product are required, each meeting the specific requirements of a local geographical region. For example, computer products for various countries may differ in the power supply module to accommodate local voltage, frequency, and plug conventions. Keyboards and manuals must match local language. Telecommunication products may also be differentiated by the communication protocols supported. In some cases, the need for localized versions of a product results from government-imposed local content requirements. It is increasingly necessary for operations managers to overcome the challenges associated with the large number of product options existing in most products.

Even within the same geographical region, there may also be multiple models of the product within a product family that relate to different functionalities or capabilities of the product. These different models may also reflect the differential needs of different market segments, e.g., business, education, personal, or government sector. Hence, it is not uncommon to have significant proliferation in product numbers within a single product family. Moreover, with rapidly changing technologies, a company often has to produce multiple versions, representing different upgrades, of the same product. These factors have contributed to the need for mass customization of a product.

The result of the globalization of the market place, diverse customer needs and rapidly changing technologies is increased proliferation of product variety. This proliferation is manifested in symptoms of various forms. First, forecasting demands for the multiple versions of a product can be a nightmare to forecasters. Forecast errors of high technology products such as computers and their peripherals can often be in excess of 400%. Forecast errors can be in the form of overforecasts for some product versions and underforecasts for others. The result is excess inventory for some products and product shortages for others. Hence, with high forecast errors, the second symptom is coexistence of high levels of inventory and simultaneously poor overall customer service. Another symptom is that, under the high pressure for customer service, many operations managers choose to simply stock as much finished goods as they can. Especially with rapid technological changes, the amount of inventory that has to be written off each year, due to obsolescence, is often in the tens or even hundreds of millions of dollars.

There are other symptoms of operational inefficiencies as a result of product variety proliferation. Since the company has to manage a larger number of product offerings, a high overhead and administrative costs are needed. Higher complexity of the product offerings would also mean higher manufacturing costs, due to more specialized processes, materials, changeovers and quality assurance methods. Since different products may require different field support materials as well as technology, it is more difficult to maintain effective general product support or customer field service provision. We refer the reader to Child et al. (1991) for a thorough discussion of the operational problems related to the complexity of product variety.

To address the above operational problems, companies can invest in resources that increase the efficiency of the supply chain (Lee and Billington, 1992). Examples of such investments include: information networks that reduce the delay of order processing information; faster means of transportation such as the use of special carriers; relocating factories that are closer to customers; more sophisticated forecasting techniques; complex supply chain inventory management systems; and various forms of factory automation for efficient material handling and processing. These measures have been successful in various degrees. In more recent years, we observe an increasing trend towards redesigning the product and process so that the negative impacts of product variety can be ameliorated. Child et al. (1991) recognized that 80% of manufacturing costs, 50% of quality, 50% of time, and business complexity can be influenced by product and process redesign. The concept is thus to redesign a production process so that the point of differentiation into the multiple products in the process is delayed as much as possible. Delaying the point of differentiation implies that the process would not commit the work-in-process into a particular finished product until a later point. This concept is known as "postponement."

When properly executed, postponement can increase the flexibility of the supply chain, enhance cost effectiveness, and improve customer service. However, postponement is not always free. In particular, the design of the product and the process can have positive or negative impact towards the implementation of postponement, and thus, it is important to consider postponement as one of the many factors in the design of the product or the process. Hence, we advocate the concept of "design for postponement."

In the next, we will describe in more detail the different levels of postponement, and the cost drivers affecting postponement. An understanding of these factors is an important step towards successful implementation of "design for postponement."

3. Postponement: Concept, Levels, and Types

In general, multiple end-products may share some common processes and/or components in the initial stages of the production process. At some point in the process, specialized processes or components are then used to customize the work-in-process, which was a generic product up to that point in the process, into the different end-products. Such a point is usually known as the point of product differentiation. Postponement refers to redesigning the process to delay the point of product differentiation. In our subsequent discussion, we refer to the *level* of postponement as the relative location of the point of product differentiation. Thus, an early level of postponement is one in which the point of product differentiation occurs early in the supply chain, whereas a late level of postponement is one in which such a point occurs late in the chain.

The concept of postponement was first introduced in the marketing literature by Alderson (1950), who defined postponement as a strategy to postpone changes in form and identity to the latest possible point in marketing. Such a concept has actually been used more extensively in the logistics and distribution side of business (see Shapiro and Heskett, 1985, Zinn and Bowersox, 1988, and Zinn, 1990). Industries in consumer products have also been using such a concept for fast response to customer orders. For example, Benetton stocks undyed garments which are dyed after the start of the selling season when more information on customer preferences can be obtained (Harvard Business School Note, 1990). Other companies in the apparel industry apparently use similar strategies (see Fisher and Raman, 1992).

Zinn (1990) defined postponement as the strategy to package, differentiate and otherwise complete a product at the latest possible point in the product flow. He defined postponement as consisting of multiple levels: time postponement, in the sense that activities that differentiate products were delayed in time as much as possible; and other activities like postponing labeling, packaging, assembly and manufacturing. These levels are more appropriate for consumer products. Our framework of postponement borrows from Zinn's, but has been extended and modified to fit the environment for the manufacturing and distribution of high technology products such as computers and peripherals.

In the manufacturing and distribution of high technology products, it is useful to use the following structure to describe the process in which product variety is proliferated. Consider the supply chain of a product consisting of five key steps: (1) manufacturing; (2) integration; (3) customization; (4) localization; and (5) packaging. The sequencing of these steps coincides with most actual cases we have observed. It also happens to correspond to, in decreasing order, the amount of process and test engineering resources needed in the processing of a product through these steps.

(a) *Manufacturing* refers to the fundamental step in which the core of the product is made. Usually, a single generic product or only a small number of products are made at this stage.

(b) *Integration* refers to the step in which the main core of the product is combined with key subassemblies. For example, in the manufacturing of printers, the printer engine and body are made at the manufacturing step, whereas the printed circuit board is integrated into the engine at the integration stage. At the integration stage, each product, when combined with different subassemblies (e.g., different circuit boards), would become different product versions.

(c) *Customization* refers to further assembly of the product with different accessories to form distinct product choices. For example, in the case of computer products, the insertions of different I/O cards, softwares, memories and accessory boards constitute different end product options.

(d) At the *localization* stage, each of the product options so far is localized to suit the local requirements of different regions or countries. For example, different countries may have different power supply requirements, different languages for the manuals, and different packaging materials. Again, we can envision that each of the product options gets multiplied into more different end products for the different regions and countries.

(e) Finally, *packaging* as a step for product variety proliferation is meaningful for some products only. For example, in computer peripherals such as toner cartridges for laser printers, the product may come in packages containing different numbers of items in them. Packaging constitutes another step for further product variety proliferation.

Figure 1 describes the product variety proliferation tree that branches out at the various stages described earlier.

In the context of the product variety proliferation structure, postponement as a strategy to defer product differentiation can be of two types: TIME postponement and FORM postponement.

Time postponement refers to delaying the various product differentiation tasks (manufacturing, integration, customization, localization and packaging) as late as possible. One obvious possibility, as indicated by Zinn (1990), is to switch from a build-to-stock to a build-to-order mode.

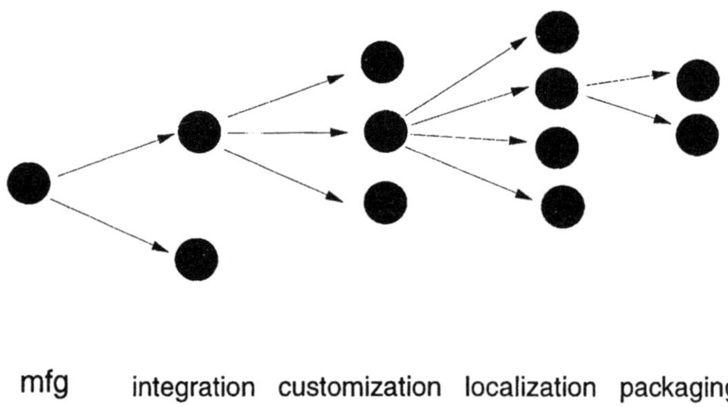

Figure 1: The Product Variety Proliferation Tree

Our experience shows that there is another great opportunity in which time postponement can be implemented. This opportunity is related to the geographical location in which the differentiating tasks are performed. In general, these tasks can be performed at the manufacturing factory, the regional distribution centers (DCs), the distribution channels such as computer dealers or resellers, and the customer sites. The earliest level of postponement refers to the case where all the tasks were performed at the factory. The latest level of postponement refers to the case where all the tasks were performed at the customer sites.

For most high technology and complex products in a global market that were traditionally manufactured and assembled at the factory, the first postponement opportunity is often to move some of the differentiation tasks to the regional DCs. Given that transit time from the factory to the DCs can take multiple weeks for ocean shipment, such a time postponement strategy effectively postpones the differentiation of the products by multiple weeks, and can result in significant savings. Lee, Billington and Carter (1992) reported such a case for Hewlett-Packard's (HP) deskjet printers, where the localization step was postponed from the factory to the DCs. Such a "design for localization" strategy has become a standard practice for that division at HP.

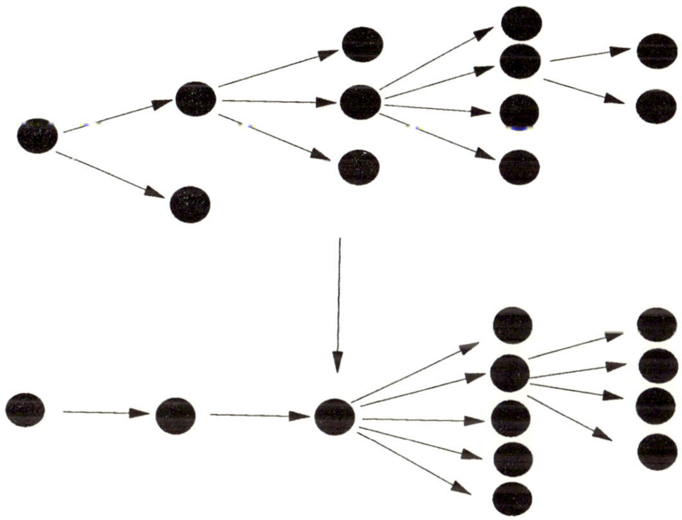

Figure 2: Form Postponement

Another type of postponement, known as form postponement, aims at standardizing the upstream stages as much as possible. Often, this is also accompanied by the standardization of components. Effectively, the point of product differentiation is then deferred. The current trend towards increased modularization and component standardization (see Ulrich, 1991, and Ulrich and Tung, 1991) makes such redesign much more feasible. For example, in the case of printer manufacturing, two key components used in the integration stage differentiate the product into monochrome versus multiple color versions. If those components can be standardized, then the two versions of the printer are not differentiated at the integration stage, thereby leading to postponement. Form postponement can thus be viewed as a process that collapses earlier branches of the product variety proliferation tree into fewer ones, as shown in Figure 2.

Design for postponement is thus a concept that calls for the redesign of the product or the process so that either time or form postponement can be achieved in a cost-effective way. Table 1 gives some examples of design for postponement principles.

Postponement Stage	Examples of Principles
Packaging	Design for bulk packing
Localization	Universal/External power supply
Customization	Modularization/Design for assembly
Integration	Slide-in integration/Modularization
Manufacturing	Standardization/Commonality

Table 1: Design for Postponement

4. Key Cost Drivers

A comprehensive analysis of the right product and process design for postponement should be carried out under a framework that considers all the relevant cost drivers impacted by postponement. Some of these drivers can be explicitly measured, while others are less tangible and can only be considered qualitatively. In this section, we discuss a list of such drivers.

4.1. Inventory. In the introduction, we described the difficulties of forecasting and the related inventory and customer service problems in the presence of product variety. Postponement should impact inventory investment and customer service. A useful means of analysis would start with an evaluation of the amount of inventory investment that would be needed with a new design that aimed at some level or form of postponement, while the same customer service target can be met. In other words, we keep the customer service level as a constant, and evaluate alternative postponement designs in terms of their associated inventory investments. The advantage of such an approach is that we can then avoid the difficult problem of comparing alternative designs that have different combinations of inventory and service. As a general rule, postponement should lead to lower inventory investment for the same service target level. The magnitude of inventory reduction, however, depends on the timing of the point of differentiation, the variability and the interdependency of the demands of the different product versions, the target service level required, and the total lead time.

The relationship between inventory investment and the above factors can be explicitly modelled analytically. We note the parallel of modelling a postponement process and of modelling a multi-echelon inventory system (e.g., Eppen and Schrage, 1981; Schwarz, 1989). In a multi-echelon inventory system with an arborescent structure, there exist multiple lower echelon sites that are descendants of an upper echelon site. This is similar to a postponement process in which multiple product versions share a common subassembly or a common prior process before the customization steps. Hence, one can borrow extensively from the existing knowledge of multi-echelon inventory theory to model postponement processes. In the case where there are no intermediate work-in-process buffer inventories, and only finished products are held in a supply chain, Lee (1992) developed a formal model that shows how inventory investment can be reduced when the point of product differentiation is deferred, i.e., a later point of postponement. That model shows that the reduction in inventory is greater when the different products are negatively correlated. Lee and Tang (1993) extended such a model to allow the stocking of intermediate work-in-process buffer inventories.

Since inventory is a central driver of postponement evaluation, we provide an illustration of this relationship. Consider the various time postponement strategies described in the earlier section. Hence, the activities related to manufacturing, integration, customization, localization and packaging can be carried out either at the factory or at the regional DCs. The transportation times between the factory and the DCs are fairly long because the bulkiness of the product requires that the product be transported by sea. This transportation time means that the more activities we allocate to the distribution location, the greater is the postponement level. Consider a particular time postponement strategy. Let X be the total time needed to complete the activities assigned to the factory, T be the transportation time from the factory to a DC, and Y be the total time needed to complete the activities assigned to the DCs. Suppose further that only finished goods inventories are held. We can view this postponement strategy as one that is equivalent to an inventory system described in Lee (1992) where the common generic process time is X+T, and the customization time is Y. Different time postponement strategies correspond to different combinations of X and Y.

Figure 3 graphically depicts inventory investments under the different time postponement strategies. As expected, the later the postponement level, the greater the inventory reduction. The interesting observation here is that the inventory reduction exhibits increasing marginal returns as we delay the level of postponement. Hence, inventory reduction is less appreciable for earlier levels of postponement, but is of greater significance for later levels of postponement.

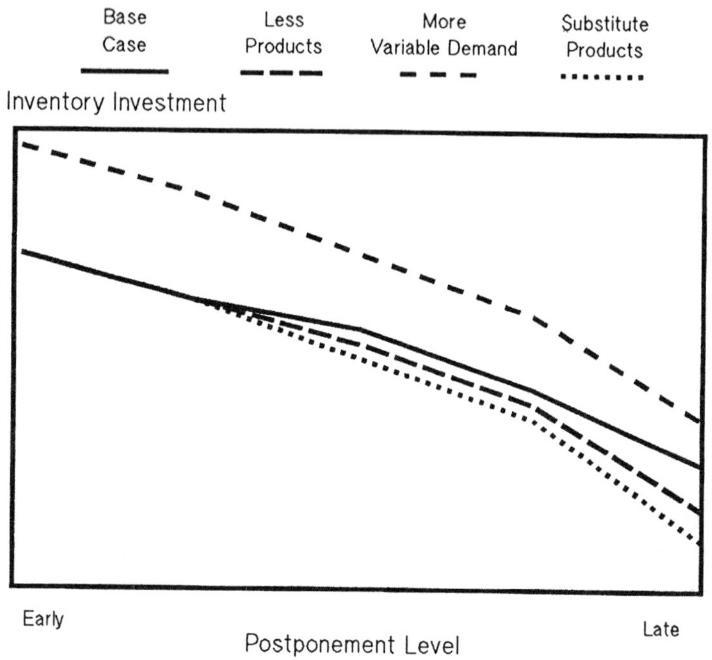

Figure 3: Inventory Investment and Postponement

Figure 3 also shows how the inventory and postponement relationship changes as a function of some key parameters. Note that postponement is more effective to help reduce inventory investments to meet the same service target when (1) the products have more variable demand; (2) the products are more negatively correlated; (3) the transportation time between the factory and the distribution location is longer; and (4) the number of product variety is larger.

4.2. Other Key Drivers. Postponement often requires some redesign of the product and/or the process, and this means that some resources are needed to make it happen. Implementation costs such as retooling, retesting, training, installation of special process equipment at distribution or downstream locations, and new design of packaging, etc., are tangible and quantifiable.

Besides one-time implementation costs, a new postponement strategy can result in different fixed and variable processing costs for the product. These costs can go up or down as the postponement level increases. For example, time postponement often requires distribution locations performing some tasks that were previously performed at a central factory. When a company has multiple distribution locations located globally, this means that some fixed overhead and equipment are now needed at multiple sites as opposed to a single one. In this case, the fixed cost component of the product cost would go up as we delay the level of time postponement. On the other hand, in the case of form postponement, we may be standardizing processes and using more and more common components. Managing fewer components means that fewer part numbers need to be monitored, and fewer suppliers need to be managed. Often, this results in lower overhead, and the fixed cost component would go down as we delay the level of form postponement. In a similar manner, the variable component of processing cost could also go up or down as we delay the level of postponement. In general, processing costs for a product have unique relationships to the level of postponement, and must be analyzed on a case by case basis.

Another key cost driver of postponement is material. For time postponement, it is possible that some of the parts that were procured centrally by the factory are now procured by the distribution locations using local sources. This could have an impact on the material cost, both in terms of the purchasing cost of the product and the inbound freight cost. For form postponement, the standardization of common parts for multiple parts can lead to higher material costs, as a standard part may contain more functionalities than needed. On the other hand, the higher purchase volumes of common parts may reduce part costs through quantity discounts and higher negotiation leverages with suppliers.

When a postponement strategy calls for rearranging processing tasks among different manufacturing and distribution locations, the nature of the items that are being transported among these sites would change. Consequently, the transportation requirements would change, leading to a change in freight costs. The cubic volume of a product will probably change. Typically, transportation costs are lower with a later level of postponement.

Another implication that is also a direct consequence of the rearrangement of processing tasks among manufacturing and distribution locations is the way products can be packaged in transportation, which can significantly affect the freight costs. Under some postponement strategies, products that are not yet fully customized and localized are being transported from one place to another. These semi-finished items can be packaged in a very compact way, i.e., bulk packing, so that pallet density can be increased. This could result in substantial freight savings. We have observed savings in freight costs of 50% under bulk packing.

In summary, there are many key cost drivers that impact the effectiveness of postponement. Inventory cost is one that can be explicitly modelled and quantified, due to existing knowledge in multi-echelon inventory theory. The impact of the other costs, however, is less clearcut.

4.3. Costs That Are Often Overlooked. Apart from the above costs, there are others that are often of great significance, but are overlooked in postponement design analysis.

In the previous section, we note that postponement could lead to the rearrangement of processing tasks among manufacturing and distribution locations, which in turn would change the nature of the items that are being transported from one place to another. In a global supply chain, these sites are often located in different countries. The change in the nature of the items could lead to different customs and duties being applied to the items that are shipped from one country to another.

Consider a time postponement strategy such that, instead of having the product fully customized and localized at the manufacturing plant, items in semi-finished form are shipped to remote distribution locations, where they are then customized and localized for specific customer needs. Products that are fully customized and localized would incur the full customs and duties, whereas items in semi-finished form may sometimes be considered to be subassemblies or even components, which incur either reduced customs and duties, or no customs and duties at all. The difference could mean a few percentage points off the total cost of the product. While customs and duties should not be the primary driver of product and process design, they certainly should be a consideration that could not be overlooked.

For high technology products, it is not uncommon for products to be returned from the customers after they are delivered. These returns are often due to minor assembly problems in the integration, customization or localization steps. A benefit of time postponement is that distribution locations that are close to the customers may now perform some customization and localization tasks, and with the necessary equipment and components, could then have the capabilities to perform minor diagnostic tests, repairs and rework of the product. This leads to much faster and responsive correction of the product returns problem. Without such postponement designs, the product may have to be shipped back to the manufacturing site for diagnosis and repair. The transportation, information and processing delays can be significant, leading to high levels of customer dissatisfaction.

On the other hand, the level of training and experience of personnel, the sophistication of equipment, and the level of engineering support at remote distribution locations may not always match those of the manufacturing plants. Hence, when the remote distribution locations are charged with some integration, customization and/or localization tasks, the risk of "unquality" may go up. Products that are of substandard quality may "slip through the crack." In other cases, more testing and rework are needed at distribution sites, leading to longer cycle times.

4.4. Factors That Are Not Readily Measurable. As discussed in the above sections, postponement may involve collaboration and analysis that would span multiple organizations within a firm: R&D, engineering, manufacturing, distribution, finance, and sales. Efforts to get organizational buy-ins and overcome organizational barriers should not be underestimated.

In the case of time postponement where distribution locations have to assume some degree of manufacturing tasks, some transformation in people's mind-set is required. For example, distribution location personnel will now have to perform manufacturing tasks such as procurement, testing and shop floor scheduling and control, which are tasks that they were not prepared for. Thus, organizational readiness is a factor that should be addressed.

On the other hand, having the remote distribution locations perform some degree of manufacturing tasks can have the added benefit of creating a manufacturing presence in these countries. With increasing protectionism and local content requirement by many local governments, such manufacturing presence could be very helpful. Moreover, manufacturing presence in a locality could also have significant marketing value. Some customers in certain countries are more attracted to products in which some manufacturing steps were performed in those countries (see Cohen and Lee, 1989).

Associated with postponement are side-benefits like ease of product upgrades and the shortening of new product development time, both of which can have significant market value. Form postponement, in particular, often requires redesigning the product with increased modular structure and part commonality. This could make product upgrades much easier. With time postponement, such upgrades could also be done at distribution locations that are much closer to customers, instead of at the manufacturing plants. Similarly, increased modular structure and part commonality also facilitate product development of future generations of the same product, since these generations often require only the redevelopment of some key modules or components.

Another benefit of having the packaging step performed at the regional DCs is the associated "green" effects. There are more and more countries that have explicit environmental regulations calling for the return of packaging materials to the country of origin, resulting in significant incremental costs to the manufacturer. By having packaging done at the regional DCs, the packaging materials would likely be locally sourced, and the problem of having to return the packaging materials to the country of origin becomes a non-problem. Also, the possibility of reusing packaging materials becomes more feasible.

5. Postponement Strategies

A postponement strategy that calls for different levels of time and form postponements would result in different cost performance with respect to the cost drivers described above. Table 2 describes the likely outcome in a structured and summarized form. A discussion of these cost drivers follows.

(1) In general, deferring differentiation tasks to distribution sites means that products with fewer components are being shipped from the factory to the DCs. In addition, it is more likely that bulk pack can be used in shipment. Hence, the freight cost is likely to be lower. When these tasks are performed at the factory, the freight cost would be higher. Next, if form postponement is implemented so that most components have been standardized, the products that are being shipped would be more alike, and the possibility of the use of bulk pack helps to reduce freight costs.

(2) Late postponement in both form and time requires the lowest level of inventory investment. When postponement is late in form but not in time, or late in time but not in form, then some moderate inventory reduction is achieved.

(3) With late time postponement where most of the differentiation tasks are performed at the regional DCs, the product being shipped would incur lower customs and duties.

(4) Late form postponement coupled with early time postponement implies that a significant portion of the manufacturing processes have been standardized, whereas manufacturing tasks are centralized at the factory. This enables economies of scale without the diseconomies of scope, and thus is the lowest processing cost strategy. On the other hand, early form postponement coupled with late time postponement implies that a large number of distinct manufacturing processes exists at multiple regional DC locations. Processing

cost is high in this case, due to diseconomies of scale with diseconomies of scope.

(5) Early form postponement coupled with late time postponement requires more distinct components to be procured at more distinct regional DC locations. The higher cost of maintaining a greater supply base and the smaller chance of quantity discount often lead to higher material cost. Highly standardized processes with common components procured centrally at a factory would probably be the lowest material cost strategy.

(6) Having more diverse manufacturing activities performed at multiple DC locations, which are not used to such activities (early form postponement with late time postponement), would constitute the biggest quality assurance and control challenge. On the other hand, when processes have been standardized and when products are primarily made centrally at the factory (early time postponement with late form postponement), the quality cost is the lowest.

Time Postponement \ Form Postponement	Early	Late
Late	▪ low freight cost ▪ moderate inventory cost ▪ low customs & duties ▪ high processing cost ▪ high material cost ▪ high quality cost	▪ low freight cost ▪ low inventory cost ▪ low customs & duties ▪ moderate processing cost ▪ moderate material cost ▪ moderate quality cost
Early	▪ high freight cost ▪ high inventory cost ▪ high customs & duties ▪ moderate processing cost ▪ moderate material cost ▪ moderate quality cost	▪ moderate freight cost ▪ moderate inventory cost ▪ high customs & duties ▪ low processing cost ▪ low material cost ▪ low quality cost

Table 2: Postponement Strategies and Costs

The above discussion indicates that different levels of time and form postponements have very different cost implications. For different products in different markets, the relative magnitudes of these cost implications could also be very different. Thus, it is important for companies to coordinate the two types of postponement and to design products and processes accordingly. For example, for products where inventory and freight costs dominate, late time and form postponement could be the effective strategy. For products with very complex manufacturing processes where processing, material and quality costs dominate, early time and late form postponement could be the effective strategy.

As an illustration, the authors have observed various postponement implementations for different products. Figure 4 shows the paths that these different products have taken in their respective redesign for postponement strategies.

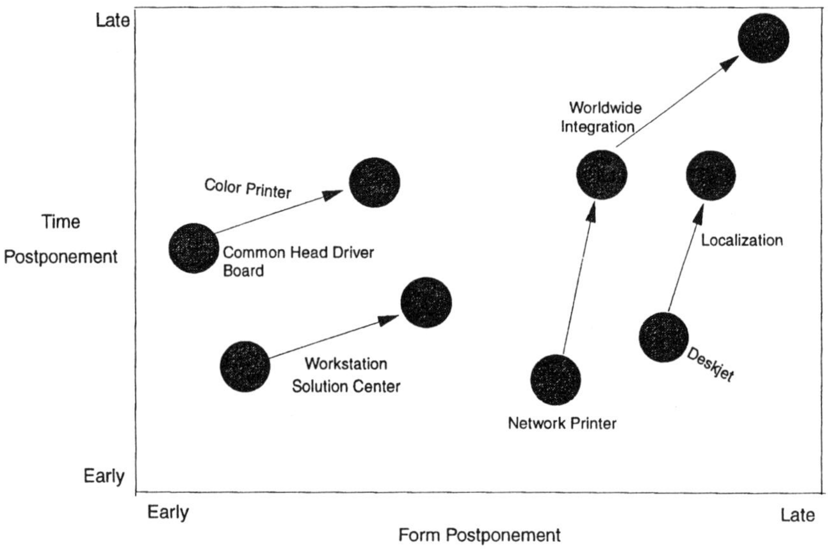

Figure 4: Examples

6. Conclusion

Postponement is an effective way to enable a company to effectively control its costs and manage service performance in the presence of increasing product variety. To exploit the full potential of postponement, we contend that the product or supply chain process should be designed to support its implementation. Design for postponement should become part of the design principles and culture. Furthermore, postponement can come in time or in form, and the two types should also be appropriately aligned for maximum benefits.

REFERENCES

Alderson, W., "Marketing Efficiency and the Principle of Postponement," *Cost and Profit Outlook*, September 1950, 3.

Child, P., R. Diederichs, F. Sanders, and S. Wisniowski, "The Management of Complexity," *Sloan Management Review*, Fall 1991, 73-80.

Cohen, M.A. and H.L. Lee, "Resource Deployment Analysis of Global Manufacturing and Distribution Network," *J. of Manufacturing and Operations Management*, 2, 1989, 81-104.

Dean, J.W. Jr. and G.I. Susman, "Organizing for Manufacturable Design," *Harvard Business Review*, January-February, 1989, 28-36.

Eppen, G.D. and L. Schrage, "Centralized Ordering Policies in a Multi-Warehouse System with Lead Times and Random Demand," in *Multi-Level Production/Inventory Systems: Theory and Practice*, edited by L.B. Schwarz, North-Holland, Amsterdam and New York, 1981.

Fisher, M., "Integrating Manufacturing, Logistics and Marketing," paper presented at the Joint National Meeting, ORSA/TIMS, San Francisco, 1992.

Fisher, M. and A. Raman, "The Value of Quick Response for Supplying Fashion Products: Analysis and Application," working paper, Department of Decision Sciences, University of Pennsylvania, 1992.

Harvard Business School Note, "Quick Response in the Apparel Industry," N9-690-038, February 27, 1990.

Lee, H.L. and C. Billington, "Pitfalls and Opportunities in Supply Chain Inventory Management," *Sloan Management Review*, 33, Spring, 1992, 65-73.

Lee, H.L., C. Billington, and B. Carter, "Hewlett-Packard Gains Control of Inventory and Service Through Design for Localization," 1992, *Interfaces*, forthcoming.

Lee, H.L., "Design for Supply Chain Management: Methods and Examples," paper presented at Conference on Perspectives in Operations Management, UCLA, November, 1991.

Lee, H.L., "Effective Management of Inventory and Service Through Product and Process Redesign," working paper, Stanford University, 1992.

Lee, H.L. and C.S. Tang, "Modelling the Costs and Benefits of Delayed Product Differentiation," working paper, UCLA, 1993.

Schwarz, L.B., "Model for Assessing the Value of Warehouse Risk-Pooling: Risk-Pooling over Outside-Supplier Leadtimes," *Management Science*, 35, 1989, 828-842.

Shapiro, R.D. and J.L. Heskett, *Logistics Strategy: Cases and Concepts*, West Publishing Company, St. Paul, Minnesota, 1985.

Taguchi, G. and D. Clausing, "Robust Quality," Harvard Business Review, January-February, 1990, 65-75.

Ulrich, K.T., "Modularity, Variety, and Standardization in Product Design," Paper presented at Production and Operations Summer Camp, M.I.T., 1991.

Ulrich, K. and K. Tung, "Fundamentals of Product Modularity," Working Paper, M.I.T., 1991.

Whitney, D.E., "Manufacturing by Design," Harvard Business Review, July-August, 1988, 83-91.

Zinn, W. and D.J. Bowersox, "Planning Physical Distribution with the Principle of Postponement," J. of Business Logistics, 9, 1988, 117-136.

Zinn, W., "Should You Assemble Products Before an Order is Received?" Business Horizons, 33, 1990, 70-73.

7

DESIGN FOR SET MANUFACTURABILITY

Reza H. Ahmadi and Hernan Wurgaft
Anderson Graduate School of Management
U.C.L.A. Los Angeles, CA 90024

Abstract: Today's competitive environment has stimulated a proliferation of products to satisfy different market segments and niches. Additionally, in order to compete effectively, firms produce several products concurrently. Concurrent product mix production reduces work-in-process and finished goods inventory, and minimizes the impact of forecast errors on the production system. Benefits of simultaneous product mix production would be marginalized because of resulting long lead times and high work-in-process, if the product set is not properly designed and managed.

In this paper, we show how manufacturers can gain a competitive edge by simultaneously considering the interaction among the product set, manufacturing system, and the operating mechanism of the production system. These interactions raise manufacturability requirements pertaining to the management of the product set. Moreover, we develop a framework, Design For Set Manufacturability, to address the different managerial issues that arise in a Flexible Manufacturing System environment.

1. Introduction

The drawbacks of the traditional, over-the-wall product development approach (c.f., Liker and Fleischer (1992)) have caused researchers and practitioners to look very closely into the product development process. The empirical evidence regarding the impact of product design on the final cost of the product (c.f., Whitney (1988)), indicates the importance of product design and development as a source of competitive advantage.

In one electronics company, engineering changes to a design "thrown over the wall" accounted for 20% of the company's total overhead costs. However, 80% of these changes were avoidable (Adler (1992)). Studies in several home appliance, automobile and aerospace companies have revealed that about 70% of the life cycle cost of a product is determined when the product is designed (c.f., Nevins and Whitney (1989)). Other items in the cost structure of a firm have a relatively smaller influence on the final cost of the product, compared with the influence of design. Figure 1 shows a typical distribution of these influences.

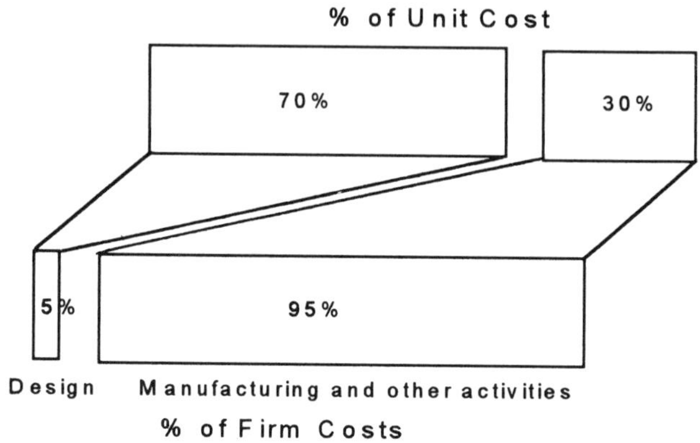

Figure 1: Influence of cost items

Design for Life-Cycle Manufacturing (DFM) is the most general conceptual category which encompasses different elements of design. Stoll (1990) defines DFM as "the full range of policies, techniques, practices, and attitudes that cause a product to be designed for the optimum manufacturing cost, the optimum achievement of manufacturing quality, and the optimum achievement of life-cycle support (serviceability, reliability, and maintainability)". Some of the better-known elements of DFM are Design for Assembly (c.f., Boothroyd and Dewhurst (1987)), Value Engineering (c.f., Salvendy (1982)), Quality Function Deployment (c.f., Hauser and Clausing (1990)), the Taguchi method (c.f., Taguchi and Clausing (1990)), and organizational schemes to attain the integration between engineering and manufacturing (c.f., Dean and Susman (1989)).

The use of DFM practices in BPC, a company that produces electrical induction motors, has led to a three to four times productivity increase and savings of 30% on the final cost of the product (c.f., Gill (1991)). FMC corporation reported that by integrating design, manufacturing and quality, they saved more than 40% on gauging cost, coordinate measuring machine programming, setup time and rework (c.f., Burget (1988)). In TI, Bogard et al.

(1991) cite estimates of benefits of 30 to 70% reduction of product development times, 200 to 600% improvement in quality, 20 to 110% increase in white collar productivity and 60 to 90 % reduction in engineering changes. Whitney (1988) asserts that a 50% reduction in the direct cost of manufacturing can typically be expected by forcing design to anticipate production requirements.

The emphasis of DFM is on defining the product design and manufacturing system design decisions so as to satisfy the manufacturability and life-cycle requirements. While in the traditional product development approach there was little interaction between design and manufacturing, DFM propounds the integration of product design and manufacturing system design to properly address the downstream requirements (see Figure 2). Optimization of the design decisions can then be carried out with respect to manufacturability and life-cycle support as well as with respect to functionality and marketability.

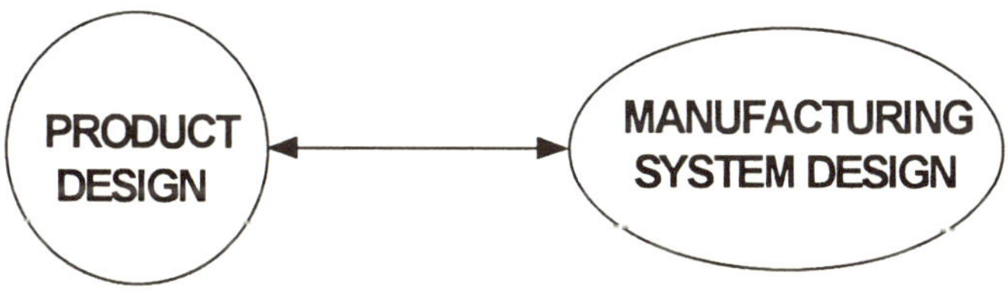

Figure 2: Design for Manufacture

Most of the work on DFM has focused on product-process development projects, where together with a new product, a new manufacturing process -and sometimes a new process technology - is being introduced. For example, Nevins and Whitney (1989) present two case studies in the automobile industry - an air conditioning module and an automobile rear axle - where the new product requires designing and building a new manufacturing shop. In their study of five successive generations of the Sony Walkman, Sanderson and Uzumeri (1992) found that teams made up of design and manufacturing engineers worked jointly to develop both the new models and the manufacturing systems to produce them.

However, in the majority of production processes, high system capital costs prohibit the design and development of a new manufacturing system each time a new product is developed and introduced. In particular, the processing capabilities of many systems, such as Flexible Manufacturing Systems (FMS's), may not be changed as a result of new product introduction. These systems are generally characterized by high degrees of automation and flexibility, and by a wide range of process capabilities. Within their basic configuration, FMS's have a number of design variables that leave latitude for usage in diverse settings. The productivity of these systems can be greatly enhanced by considering the manufacturing system design variables as well as the product design decisions. Furthermore, batch manufacturing accounts for 80% of the production volume in the United States.

Up to now, most of the design guidelines, rules, and constraints which have been developed only consider the production of a single product in isolation. But many firms, in order to compete effectively, produce a set of products concurrently. Concurrent product mix production reduces work-in-process and finished goods inventory, and minimizes the impact of forecast errors on the production system. The simultaneous production of a set of products gives rise to product set manufacturability requirements. When the set of products is not properly managed, products have long manufacturing lead times and spend large amounts of time in queues as work-in-process, and the system's throughput rate is low (in the metal-cutting industry, the average processing time of a workpiece is only 5% of its manufacturing throughput time (c.f., Rembold et al.(1985))). These observations motivate the development of our framework referred to as Design For Set Manufacturability (DFSM). DFSM considers the interaction and implication of the production of a set of products for the optimal performance of the manufacturing system. Figure 3 represents the elements of DFSM. Here, in addition to the product and manufacturing system design decisions, the operating conditions for the set of products have to be considered at the design level.

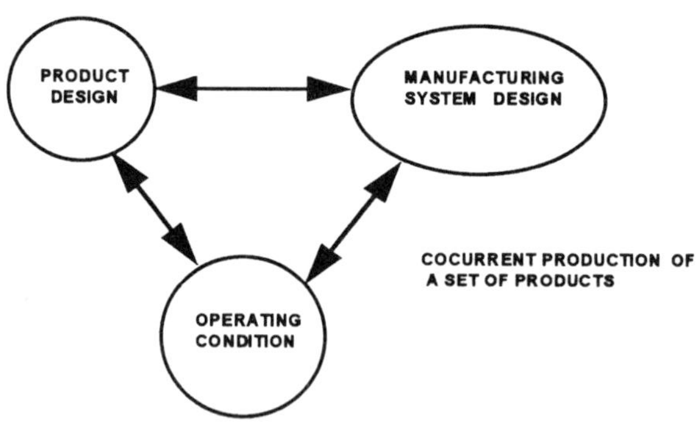

Figure 3: Design for Set Manufacturability

The rest of the paper is organized as follows. In section 2 we characterize a simple operating condition for concurrent product mix production that enhances the performance of the manufacturing system. Section 3 describes the manufacturing system design considerations. In section 4 we describe the product design representation. Section 5 illustrates how the product design and manufacturing system design can be used to satisfy the product set process consistency requirements. In section 6 we develop the conceptual methodology to optimize the system's throughput rate. In section 7 we provide insights based on DFSM for several major managerial issues. The conclusion and summary of the paper are presented in section 8.

2. Simultaneous Production

In this section we describe the ideal operating condition to manufacture the set of products. Flexible manufacturing systems represent large investments, and consequently the effective utilization of these systems is quite important (c.f. Jaikumar (1986)). The successful management of such systems implies short manufacturing lead times and operating under low work-in-process conditions. These characteristics can best be illustrated by considering the operating condition of a perfectly balanced assembly line, where products move smoothly and continuously from one station to the next with no waiting in queues. Figure 4 depicts the linear flow under this environment. The system capacity is fully utilized, the throughput rate is maximized, and the system has fast response time and operates with minimal work-in-process. We refer to this ideal operating condition as synchronized flow manufacturing.

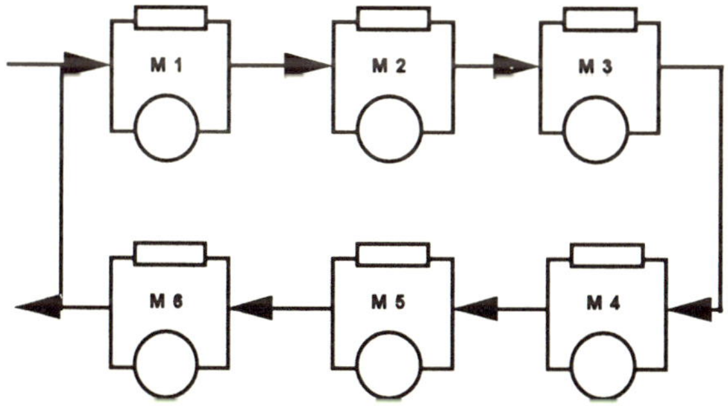

Figure 4: Synchronized Flow

To capture the benefits of synchronized flow, firms commonly produce several products concurrently. Simultaneous production can best be carried out by repetitively producing a Minimal Part Set (MPS), which is the smallest set of products that satisfy the demand ratio for the total product demand. Associated with an MPS is a cyclic or periodic schedule, in which the units of each MPS go through the system in exactly the same order. The periodic schedule has a timing pattern for the starting time of the operations, that repeats itself for each MPS (c.f. Hitz (1980)). The repetitive production of an MPS has strong benefits over the sequential production of large batches. It achieves timely delivery and reduced inventory, when there is a constant proportion between the demand rates of products. It increases the system's responsiveness to relative changes on the demand rate because as these changes occur, the composition of the MPS can be modified accordingly. Finally, since periodic scheduling repeats an identical mixture of items, the production is more predictable, and scheduling and flow control are simplified (c.f., Stecke and Kim (1992)).

The synchronized flow operating condition contrasts with the job shop operating mechanism that is commonly used to manage the production flow in advanced manufacturing systems (see Figure 5). In the presence of contention for bottleneck machines or for the material handling system, products spend large amount of time waiting in the queues, lead times are long and the throughput rates are low. Empirical evidence shows that parts may spend over 90% of their time in a job shop waiting in queues (c.f. Karmarkar et. al. (1985)). A deliberate design effort has to be made to avoid the complexities of having to manage a system as a job shop and to attain a synchronized flow, when different products with different processing requirements and routings are produced.

The periodic scheduling of an MPS is propitious for implementing a synchronized flow. However, two conditions have to be satisfied to make the synchronized flow possible. These conditions describe the set manufacturability requirements within the DFSM framework. The first condition is that the products in the MPS should admit the same linear flow through the production system. In this way, they can be moved and processed as a set. This constraint is referred to as product set process consistency. This constraint can be satisfied both at the product design and manufacturing system design activity levels. Consequently, the product design activity becomes rather a product set design effort, with an emphasis on selecting a set of products to be simultaneously manufactured and making it process consistent. The second condition for synchronized flow is that the workload of the stations has to be approximately balanced. Balancing the workload across production stations maximizes the throughput rate and optimizes the performance of the production system. Again, significant performance improvements can be achieved by considering this condition at the product design level.

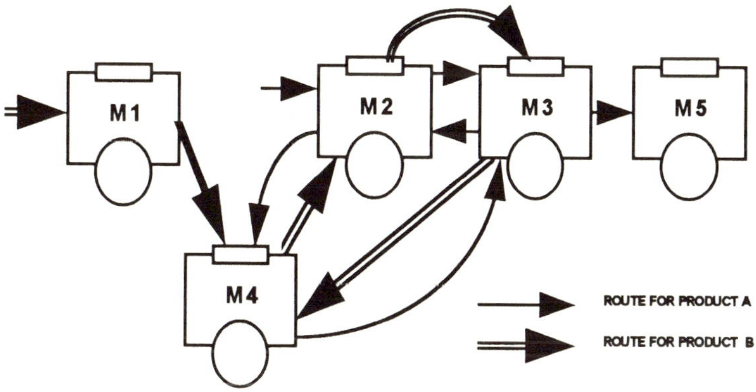

Figure 5: Job Shop Arrangement

In this section we have presented the synchronized flow system constraints and its associated set manufacturability requirements. The product set process consistency requirement stems from a feasibility condition for synchronized flow. The workload balancing requirement tends to optimize the performance of the synchronized flow system.

3. Manufacturing System Design

In this section we shall present the manufacturing system design decisions that can be used to influence the flow of the product set. We shall concentrate our discussion on FMS's that make metal parts from castings or formed stock by means of metal cutting processes. These fabrication systems are characterized by high degrees of automation and flexibility that result from the use of automated material handling devices and Computer Numerically Controlled (CNC) machine tools. The design variables that we shall identify for this manufacturing environment can be adapted to other advanced manufacturing systems as well.

A typical FMS consists of a set of Horizontal Machining Centers (HMC). Four and five axis CNC machining centers can perform complex milling, boring, drilling and tapping operations without necessitating part refixturing. They combine the operations often performed on a series of milling machines, drill presses and other specialized machines (c.f., Stecke (1983)). The variety of workpieces that an FMS can simultaneously produce in an unmanned fashion is mainly limited by the tooling capacity of the machines. Modern HMC's used in FMS have tool carrousels with provision for 50 to 150 tools (c.f., Talavage and Hannam (1988)). Part of the tooling capacity is normally used to accommodate `sister' toolings, which are duplicates of either fragile tools or tools that wear rapidly. Even after considering sister tooling, the capacity of modern HMC's allows great flexibility for processing a variety of parts within certain size, material and geometrical tolerance range.

The choice of the material handling system and the layout are two important and related FMS design decisions. They are made at the time of building the system, usually in agreement with a synchronized flow. A synchronized flow supposes a unidirectional loop layout arrangement that connects all workstations by a path passing through each workstation exactly once. Unicyclic material handling systems, like loop conveyors, tow lines, overhead monorail systems or wire paths of a unidirectional automated guided vehicle satisfy this requirement of a synchronized flow system. Also, these systems are usually preferred in practice due to their relatively lower initial investment cost (c.f., Afentakis (1989)). Thus, for most FMS's, the basic layout and material handling system configuration allows the implementation of a synchronized flow.

The unmanned operation of an FMS gives rise to a setup problem that consists of the loading of the tool carousels with tools. Since the production flow will be greatly determined by the way operations are assigned to the machines, we consider that the loading of the tool carousels is a fundamental manufacturing system design variable. Each cutting tool can be associated with one or more operations required by the set of parts to be produced simultaneously. After the tools are loaded, the identical HMC's are differentiated, and it is possible to define a matrix specifying the operations that each machine can perform. We refer to the design variable given by the loading of tools as system staging. In a more general environment, system staging refers to the assignment of operations to machines, subject to technological and capacity constraints. System staging is a design decision that has to be considered every time a change on the product mix takes place.

Other manufacturing system design variables are related to system capacity configuration. These variables involve deciding on the number of machines and the grouping of machines into workstations. The total number of machines is usually fixed at the time of constructing the manufacturing system. But machines are normally grouped into workstations with the purpose of balancing the workload, managing bottlenecks, and optimizing the performance of the system. Machine grouping is a variable to be actively used to optimize the performance of the synchronized flow.

The manufacturing system design decisions that affect the set manufacturability requirements in metal-cutting FMS's are: layout decisions, material handling system selection, machine configurations and system capacity configuration decisions. Layout and material handling systems can be easily made consistent with the requirements for synchronized flow. Machine staging and system capacity configuration can be used as decision variables to satisfy the set manufacturability requirements. As the characteristics and mix of the product set change, these configurations can be adjusted accordingly to maximize the system performance.

4. Product Design

In this section we shall present the product design decisions that can be used to satisfy the set manufacturability requirements. We shall introduce a representation of alternative product designs that allows a better integration between engineering design and manufacturing to attain a synchronized flow.

Engineering designers, in the design process, develop a description of the product that consists of a set of engineering drawings and other relevant engineering data specifying the form of the product in terms of geometric information, tolerance and material types. This information is typically supported by engineering prototypes, engineering analysis, and CAD files. In the traditional product development approach, most of the input used by designers comes from marketing. This input describes the functionality of the product as target needs, problems that the product solves, and user expectations and desires. DFM also attempts to incorporate input from manufacturing, in the form of manufacturability information (c.f., Stoll (1990)).

Engineering designers normally narrow the product design to a single final engineering description. This narrowing process involves many economic and technical decisions that require a great deal of information. The quality of the decisions depends directly on the completeness, correctness, and availability of needed information. The lack of detailed manufacturability information causes the narrowing process to be suboptimal. Further optimization would be possible if the detailed set manufacturability information were integrated with the engineering design. In order to make the integration possible, we propose that engineering designers should provide a set of alternative descriptions satisfying the functional requirements. This would allow manufacturing to optimize the final choice by using all the detailed manufacturing information. Ulrich (1992) referres to the mapping from functional elements to physical components in the design process by showing how the same functional requirement can be mapped to alternative components and designs. For example, in the fabrication of the cylinder heads of small diesel engines, various designs can be adopted without major effect on the port flows and swirls (c.f., Stevens and Tawil (1986)).

The structure of design problems naturally leads to the generation of alternative designs that can be described by the use of AND-OR graphs (c.f., Eastman (1988)). On an AND-OR graph, the AND nodes define the technologies selected for use and their necessary components. The OR nodes are the local alternative realizations of the various technologies. An abstract view of the design process involves defining the AND-OR graph. The definition of such a graph involves defining the AND nodes and searching among the relevant OR branches for a combination that best meets defined goals.

A proper representation for the engineering alternatives must enable manufacturing to select an optimal design description for each product. Rather than geometric information and engineering data, a proper product representation for manufacturing is given by additional alternative sets of operations to fabricate the parts and products. Such a representation facilitates the consideration of the production flow set manufacturability conditions. The level of description of the operations specifies the tooling, cutting positions, processing times and other technical data, but it does not specify the machine where they are to be performed. The alternative sets of operations for a same engineering description, sometimes referred to as alternative process plans (c.f., Rembold et al. (1985)), may be used to represent the product manufacturing information. For example, a slot can be milled in one pass using a cutter of a certain width, or multiple passes of a smaller width.

Lin and Solberg (1991) propose a graph representation for alternative process plans that also follows the AND-OR graph of the design process. We extend this representation, which we shall call the Design Graph, to capture alternative engineering designs as well as alternative process plans for each design (see Figure 6). On a Design Graph, the nodes represent the operations required by the alternative designs. Duplication of nodes is allowed when needed. The nodes can also store other information such as resource requirements, tools and components. The directed arcs in the graph represent the physical precedence relations between the operations. An AND node in the graph means that all of its immediate succeeding operations must be performed. An OR node in the graph means that exactly one of the immediate succeeding operations needs to be performed. A node is called solved if it has no successors, the node is an AND node and all of its immediate successors are solved, or the node is an OR node and exactly one of its immediate successors is solved. A solution to an AND-OR graph is given by a subgraph that is sufficient to show that the initial node is solved. Each solution to the AND-OR graph represents an alternative product design in terms of the operations that the design requires. Manufacturing managers can choose among these alternatives to satisfy the set manufacturability conditions.

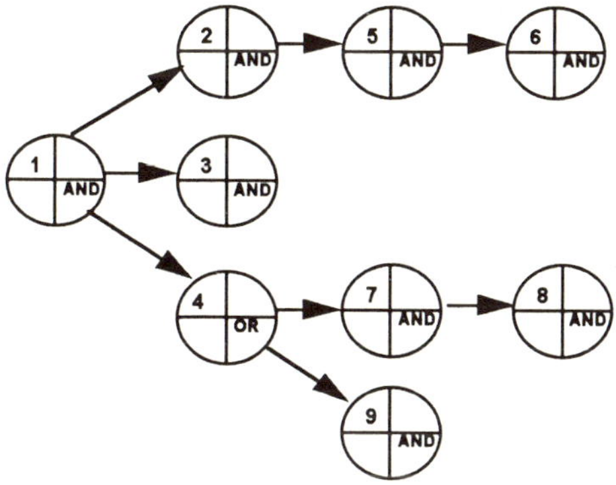

Figure 6: Design Graph

We have presented a representation of alternative product designs that follows the AND-OR structure of design problems and is based on the operations required to produce the designs. In the next sections we shall see how the alternative product designs, as well as the manufacturing system design decisions, influence the set manufacturability requirements.

5. Design for product set process consistency

The Design Graph offers the opportunity to model the characteristic of the set of products with respect to the process consistency requirement. A particular design for a product can be obtained as a solution in the Design Graph. Each solution can be represented as an acyclic digraph $G_p=(N_p,A_p)$, where N_p is the set of operations to complete product p and A_p represents the precedence relations on these operations. A representation of the flow of the set of products can be constructed by superimposing the acyclic digraphs of all the products to be simultaneously produced. The superimposed graph $G=U_{p\epsilon P}G_p$ allows us to verify if all the parts in the set admit a same linear flow through the operations. If G is acyclic, then the product set is process consistent, since all parts can be processed in a single pass through the production system. In case G contains circuits, products have to be processed individually or in groups which are process consistent. Layout and material handling scheduling become quite complex problems to manage.

With the proliferation of products resulting from the emphasis on product development and product variety as a source of competitive advantage, it is likely that sets of products will be process inconsistent unless a deliberate design effort is made to attain an acyclic superimposed graph G. In attaining process consistency, the interaction between the products is essential, and the design activity has to be thought of as product set design. Once the process consistency has been recognized as a manufacturability requirement, several set design and manufacturing system design approaches can be adopted to satisfy the requirement. The descriptions of three approaches follow.

The first approach consists of partitioning the set of products. We refer to this problem as a product set design problem. This approach utilizes the notions of Plant-within-Plant (PWP) and Focused Factory introduced by Skinner (1974). While a subset without cycles of the set of products is to be produced using a synchronized flow, the remainder subset has to be produced in a different facility or considered as a candidate for subcontracting. The partitioning can be performed by verifying if the set of products can be partitioned into two groups, each of them without cycles, or by seeking the largest subset of products that does not have any cycle (c.f., Ahmadi and Wurgaft (1992)).

The second approach is a manufacturing system design approach. The idea is to use additional machines or staging capacity slack to remove cycles from the superimposed graph G. This is a manufacturing system design approach to satisfy the set manufacturability requirement. The excess staging capacity can be used to remove directed circuits by replicating the assignment of some of the operations. The replicated operations become different nodes in the superimposed graph, and in that way the cycles are removed. Since the staging capacity is limited, this decyclization is performed at a minimum replication expense (c.f., Ahmadi and Wurgaft (1992)).

The third is a product design approach. It seeks to choose a design for each product so that the resulting set will be process consistent. This approach makes use of the alternative designs represented in the Design Graph, and optimizes the final design at the manufacturing level to satisfy the set manufacturability requirement. Figure 7 illustrates this process with a simple example. There are two products, P1 and P2, and product P1 has two alternative designs. We assume that the staging capacity is equal to 1, so each operation has to be assigned to a different machine. If the design for product P1 is chosen, then the superimposed graph has a circuit and it is not compatible with a synchronized flow. If, in turn, the other design is chosen, the resulting superimposed graph is acyclic, and the two products become process consistent. The example shows how the narrowing down of the designs without taking into account the set manufacturability information can lead to undesirable results. The concept of Design for Set Manufacturability provides a framework to choose product designs that consider the interaction between the products and yield the best operating condition.

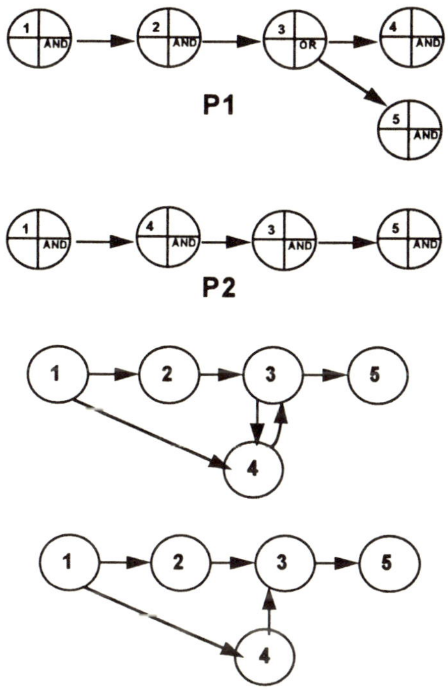

Figure 7: Product Design for Process Consistency

6. Design for workload balancing

The second set manufacturability requirement for synchronized flow, once the product set has been made process consistent, is to balance the workload of the workstation. The purpose of balancing the workload is to optimize the performance of the synchronized flow by minimizing the cycle time, and hence maximizing the throughput rate. In this section we shall discuss how both, manufacturing system design variables and product design variables can be used to satisfy the workload balancing manufacturability condition.

Workload balancing is a common manufacturing system design criterion. In the traditional problem of designing an assembly line, operations are assigned to stations and capacity is allocated so as to minimize the maximum workload, which defines the cycle time of the line (c.f., Baybars (1986)). In metal-cutting FMS's, workload balancing has also been a criterion to load the machines with tools. However, in the absence of a simple operating condition, workload balancing does not guarantee the optimization of the performance of the system (c.f., Stecke and Solberg (1984)). Under the synchronized flow condition, the behavior of an advanced manufacturing system is analogous to that of a single product assembly line. The maximum workload defines the cycle time and throughput rate of the system. The assignment of operations and allocation of manufacturing capacity has to be performed to minimize the maximum workload, also taking into account the constraint given by the staging capacity (c.f., Ahmadi and Wurgaft (1992)). In this manner, the optimization of the performance of the system is guaranteed.

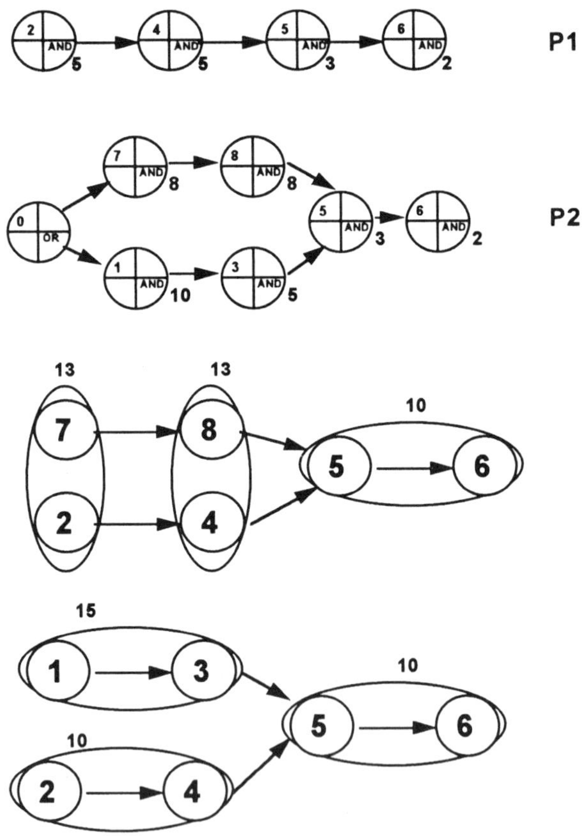

Figure 8: Product Design for Workload Balancing

The choice of the product design can also be used to affect the workload balancing and the throughput rate. This is shown in the example in Figure 8. There are two products, P1 and P2, where product P2 has two alternative designs. Whatever design is chosen, six operations have to be assigned to three machines, each with a staging capacity of two. The optimal operations assignment is shown for each of the alternative designs. If the design with the shortest processing time is chosen, then the optimal assignment yields a maximum workload of 15, and thus a throughput rate of 1/15. If, in turn, the design with a longer processing time is chosen, the optimal operations assignment yields a maximum workload of only 13, and thus a better throughput rate of 1/13. The choice of the alternative with the shortest processing time seems to be reasonable at the design stage, but it ignores the interaction with the other products and the operating condition of the shop. Superior results can be obtained, in terms of productivity improvement, if the product design decisions are made taking into account the set manufacturability requirements.

7. DFSM and managerial Concerns

The Design for Set Manufacturability framework brings about a new dimension to a host of managerial decisions. The notions of set manufacturability requirements and set design to satisfy these requirements provide a new angle for considering these decisions.

7.1. New Product Introduction. Product variety has become an important element of manufacturing competitiveness, as companies attempt to attract customers by increasing customization and offering solutions for specific functional requirements (c.f., Kekre and Srinivasan (1990)). Product variety supposes a continuous introduction of new products to satisfy ever more demanding customer needs. The new products are often derivatives of existing products, or they belong to the same platform as the existing products. Derivatives are cost-reduced versions, add-ons or enhancements of existing products for an existing production process. Products in a platform entail more changes than derivatives do, but they do not introduce new technologies or materials (c.f., Wheelwright and Clark (1992)). It is straightforward to conceive the new derivative or platform products as elements of the product set. The characteristics of the product set provide guidelines for the design of the new product. In particular, the guidelines regarding the production flow set manufacturability requirement are given by the superimposed graph of the current set of products. The operations for the new product are to be chosen from the superimposed graph, as well as the order relations. The generation of new product design guidelines is also related to the concept of virtual design, which is defined as the capability to store information about product design and function and to reuse it in designing new products (c.f., Sanderson (1989)). The use of such a capability not only results in a better manufacturing system performance, but also in a faster and more efficient new product design activity.

7.2. Product Redesign. Several factors cause a product to be renewed and redesigned during its life cycle (c.f., Shirley (1990)). Technological change can reduce the cost to satisfy the same functional requirements. Also, new functional elements can be added or fine-tuned for specific customers. These types of product change are closely related to the previous discussion on product variety and new product introduction. Another cause for product redesign is related to the set manufacturability requirements. The production volume of each product in the set will vary with its life-cycle stage. As the volume of a product grows, its influence on the synchronized flow performance will be greater. With the changes on the production rates, a different alternative from the design graph will yield an overall better manufacturing system performance. This type of redesign we refer to as set manufacturability redesign. While the functionality of the products remain the same, their set manufacturability will change with the redesign. The guidelines for the design of new products will also evolve with the redesign of the existing products.

Redesign is also important to attain product set process consistency. The example shown in section 5 is very simple, but when the set of products to be produced concurrently is large, the process of choosing a design for each product so that the set is process consistent is difficult. Sometimes, it might be necessary to generate new alternatives, not represented in the Design Graph of a product. An iterative procedure to choose new alternatives from the Design Graph or to generate new alternatives uses the superimposed graph as redesign guidelines. The process consistency requirement is to be satisfied at a minimum redesign effort. Suppose that the designs are chosen so that the first superimposed graph has cycles. These cycles provide information about what products to redesign. It is desirable to remove the cycles at a minimum redesign effort. The superimposed graph and the Design Graph of each product can be used as analytical tools to perform the redesign to satisfy the process consistency manufacturability requirement.

7.3. Cost of Variety. While product variety has become an important competitive weapon, the proliferation of products does not take place without a cost. Greater variety implies greater complexity that has a cost implication. For every doubling of variety, the typical increase of the unit cost is 20 to 35 percent (c.f.,

Mather (1992)). A high variety can be produced by any system at some cost, and the challenge is to produce the desired variety economically. The capability to produce variety economically is usually associated with flexibility. The investment in flexible technology and flexible process equipment such as CNC machines is assessed with respect to the savings in the fixed tooling and switchover costs of producing variety (c.f., Stalk (1988)). Similarly, the cost of design and redesign within the Design for Set Manufacturability framework should be assessed with respect to the reduction of the costs of variety. In this case, the relevant costs of variety have to do with the costs of products waiting,longer manufacturing lead times, and the opportunity costs of inefficient capacity utilization and suboptimal throughput rate.

7.4. Product grouping and family formation. Group Technology coding and classification techniques have a widespread acceptance in industry. These techniques exploit the sameness or similarity of parts by grouping them based on their geometrical shape and similarities in their production process. (c.f., Burbidge (1975)). The Group Technology criteria have a fundamental role to play in advanced manufacturing systems. For example, a broad classification of metal-cutting FMS's distinguishes those systems that produce cylindrical parts from those that produce prismatic parts (c.f., Talavage and Hannan (1988)). However, Group Technology information is insufficient to optimize the performance of advanced manufacturing systems because Group Technology has not been devised to consider the product flow in the presence of high degrees of flexibility. Product grouping in flexible systems was considered by Ahmadi and Tang (1991), where the product flow was indirectly addressed by minimizing the workload of the material handling system. A job shop arrangement for the system was assumed. The Design for Set Manufacturability framework provides a highly desirable criterion for family formation in the presence of high degrees of flexibility. It takes into account more complete information about the production flow and manufacturing system capabilities so that, by satisfying the set manufacturability requirements, a simple operating condition leading to enhanced productivity can be attained.

7.5. Make-or-buy decision. The Design for Set Manufacturability criteria for product grouping and family formation also apply to the make-or-buy decision. Products that are not process compatible with the set have to be subcontracted or produced by another shop within the same facility. Also, products may not be completely produced in-house or completely subcontracted. Another method to partition the operations of a product set defines alternative criteria for make-or-buy decisions. One such scheme is partitioning the operations across their processing routes such that the in-house portion of the operations yields an acyclic graph. The set manufacturability criteria has to be combined with other standard cost criteria for the make-or-buy or subcontracting decision.

7.6. Quality. The use of the design graph allows us to capture quality considerations in the generation of the design alternatives. For example, in the context of assembly, each node that represents an assembly operation can be associated with a probability that the assembly is defect free. The computation of this probability is based on a quality control effort measure, the assembly operation time, the number of assembly operations and the probability that parts are defect free (c.f., Barcan and Hinckley (1992)). Each node and each path in the Design Graph can be associated to a quality measure. The design choices, and their effect in terms of throughput rate and redesign effort, can be assessed with respect to the quality consideration.

8. Summary

A main emphasis in Design for Manufacture is to integrate downstream requirements at the product design activity levels. In advanced manufacturing systems, the simultaneous production of a set of products supports strategies for competitive advantage that compete in multiple market segments and continuously introduce new products. This observation motivates a new set of manufacturability requirements which are components of DFSM.

We characterize an ideal operating condition for the shop that is based on the synchronized flow of a minimal part set. The synchronized flow achieves short manufacturing lead times, little waiting of parts in queues, high overall utilization and a high throughput rate. Two conditions, which are stated as the manufacturability requirements within the Design for Set Manufacturability framework, have to be satisfied to make the synchronized flow possible. These conditions are product set process consistency - that all the parts in the minimal part set admit the same linear flow through the operations -and workload balancing.

Both manufacturing system design and product design decisions can be used to satisfy the manufacturability requirements. The manufacturing system design decisions are related to the configuration of the system, the grouping of machines into workstations, and the assignment of operations to the machines. The product design decisions are related to choosing a sequence of operations to produce the products. A Design Graph to represent alternative product designs is proposed. The Design Graph represents both alternative product designs and alternative process plans to manufacture a particular design. We described how the design decisions influence the set manufacturability and proposed tools for decision making. Although we focus our discussion on metal-cutting FMS, the design decisions and set manufacturability framework can be adapted to other environments like printed circuit board assembly, where operations consist of inserting or mounting components on an electronic board. Finally, we examined a set of managerial issues, including new product introduction, redesign, product grouping and family formation, make-or-buy decision, and quality, in the light of the framework of Design for Set Manufacturability.

REFERENCES

Adler, P.S., "Managing DFM: Learning to Coordinate Product and Process Design," in Integrating Design and Manufacturing for Competitive Advantage, ed. G.I. Susman, Oxford University Press, 1992.

Afentakis, P, "A loop layout design problem for flexible manufacturing systems," The International Journal of Flexible Manufacturing Systems, Vol. 1, 1989.

Ahmadi, R.H. and H. Wurgaft, "Design for Synchronized Flow Manufacturing," Working Paper, Anderson Graduate School of Management at UCLA, 1992.

Ahmadi, R.H. and C.S. Tang, "An Operation Partitioning Problem for Automated System Design," Operations Research Vol. 39, No. 5, pp.824-835, 1991.

Barkan, P. and M. Hinckley, "Benefits and Limitations of Structured Methodologies in Product Design," Conference on Design Management, UCLA, 1992.

Baybars, I., "A Survey of Exact Algorithms for the Simple Assembly Line Balancing Problem," Management Science, Vol. 32, 1986.

Bogard, T., B. Hawisczak and T. Monro, "Concurrent Engineering environments: Texas Instruments Implements Design for Manufacture," Printed Circuit Design, 30-43, 1991.

Boothroyd G., and P. Dewhurst, "Product Design for Assembly," _Boothroyd and Dewhurst Inc., Wakefield, R.I., 1987.

Burbidge, J.L., The Introduction of Group Technology, John Wiley and Sons, 1975.

Burget, P., "FMC integrates design and production; cuts costs 40%," Manufacturing Weekly, February 8, 1988.

Dean,Jr. J.W. and G.I. Susman, "Organizing for Manufacturable Design," Harvard Business Review, January-February, 1989.

Eastman, C., "Automatic Composition of Design," in Design Theory, eds. S. Newsome et al., New york, Springer-Verlag, 1988.

Gill, H., "Design for manufacture - a case study," in Design For Manufacture: Strategies, Principles and Techniques, eds., J. Corbett et al., Addison-Wesley, 1991.

Hauser J.R. and D. Clausing, "The House of Quality," Harvard Business Review, May-June 1988.

Hitz, K.L., "Scheduling of flow-shops II," Technical Report LIDS-R-1049, Laboratory for Information and Decision Systems, MIT, Cambridge, MA, 1980.

Jaikumar, R., "Postindustrial Manufacturing," Harvard Business Review, Vol. 64, No. 6, 1986.

Karmarkar, U.S., S. Kekre and S. Kekre, "Lotsizing in multi-item multi-machine job shops," IIE Transactions, Vol. 17, No. 3, 1985.

Kekre, S. and K. Srinivasan, "Broader product line: a necessity to achieve success?," Management Science, Vol. 36, No. 10, 1990.

Liker, J.K., and M. Fleischer, "Organizational Context Barriers to DFM," in Integrating Design and Manufacturing for Competitive Advantage, ed. G.I. Susman, Oxford University Press, 1992.

Lin, G. Y-J., and J.J. Solberg, "Effectiveness of Flexible Routing," The International Journal of Flexible Manufacturing Systems, Vol 3, pp. 89-211, 1991.

Mather, H., "Optimize your Product Variety," Production and Inventory Management Journal, Second Quarter, pp.38-42, 1992.

Nevins, J.L., and D.E. Whitney, Concurrent Design of Products & Processes, McGraw-Hill, 1989.

Rembold, U., C. Blume and R. Dillman, Computer-Integrated Manufacturing Technology and Systems, Marcel Dekker, 1985.

Salvendy, G. (ed.): "Value Engineering," in Handbook of Industrial Engineering, John Wiley, New York, 1982.

Sanderson, S.W., "Strategies for New Product Design and Product Renewal," Academy of Management Best Papers Proceedings, pp. 296-300,1989.

Sanderson, S.W. and V. Uzumeri, "Industrial Design: The Leading Edge of Product Development for World Markets," Design Management Journal, Vol. 3, No. 2, 1992.

Shirley, G.V., "Models for Managing the Redesign and Manufacture of Product Sets," Journal of Manufacturing and Operations Management, Vol.3, No. 2, 1990.

Skinner, W.,"The Focused Factory," Harvard Business Review, pp. 111-122, 1974.

Stalk, G., "Time, the next source of competitive advantage," Harvard Business Review, Vol. 66, No. 4, 1988.

Stecke, K.E., "Formulation and solution of nonlinear integer production problems for flexible manufacturing systems," Management Science, Vol. 29, No. 3, 1983.

Stecke, K.E., and J.J. Solberg, "The optimality of unbalanced workloads and machine group sizes in closed queuing networks of multiserver queues," Operations Research, Vol. 33, 1984.

Stecke, K.E. and I.Kim, "A flexible approach to part type selection in flexible flow systems using part mix ratios," International Journal of Production Research, Vol. 29, No. 1.

Stoll, H.W., "Design for Life-Cycle Manufacturing," in Managing the Design-Manufacturing Process, eds. J.E. Ettlie and H.W. Stoll, McGraw-Hill, 1990.

Stevens, J.L. and N. Tawil, "A new range of industrial diesel engines with emphasis on cost-reduction techniques," in Design for Manufacture,strategies, principles and techniques, J. Corbett, M. Dooner, J. Meleka and C. Pym editors, Addison Wesley, 1991.

Talavage and Hannan, Flexible Manufacturing Systems in Practice: Applications, Design and Simulation, Marcel Dekker, 1988.

Taguchi G. and D. Clausing, "Robust Quality," Harvard Business Review, January-February, 1990.

Ulrich, K.T., "The Role of Product Architecture in the Manufacturing Firm," M.I.T. Sloan School Working Paper #3483-92-MSA, 1992.

Wheelwright S.C. and K.B. Clark, "Creating Project Plans to Focus," Harvard Business Review, March-April 1992.

Wittrock, R.J., "Scheduling Algorithms for flexible flow lines," IBM Journal of Research and Development, Vol. 29, 1985.

Whitney, D.E, "Manufacturing by Design," Harvard Business Review, July-August 1988.

8

DESIGN SUPPORT FOR CONCURRENT ENGINEERING: A CASE EXAMPLE

Jayant S. Karmarkar
Innovative Configurations Inc.

Uday S. Karmarkar
Wm. E. Simon Graduate School of Business Administration
University of Rochester
Rochester, New York 14627

Acknowledgement: Discussions with P. Srinivasan and V. Marathe were very valuable in the preparation of this paper.

Abstract: Concurrent engineering requires the integration of several layers of decision, design and evaluation processes. The design decisions can include functional specifications, technology selection, functional design, physical design and component selection. Performance measures for design range from functional performance and physical measures (volume and weight) to operational issues such as manufacturability and field serviceability. This paper describes a framework for concurrent engineering with application to an area of digital electronic equipment design. Issues of design automation, data management and the incorporation of search methods are discussed and the current needs for tools in a specific area of electronic equipment design are described.

1. Introduction

The term "concurrent engineering" refers to the integration of engineering design activities to include all aspects of the process, from the functional conceptualization of the product through functional and physical design, to evaluation of the manufacturability and usability of the design. In this paper we develop a framework for the automation and support of concurrent engineering through software tools, based on the specific case of the design of multi-chip modules in digital electronic equipment.

Recent interest in concurrent engineering and engineering design, together with the automation or support of these activities, is due to several factors. First, the activity is expensive, depending as it does on skilled labor inputs. Furthermore, the design process has significant leverage on costs that are incurred subsequently. For example, it has been estimated that while only 10-15% of the unit cost of a product may be attributable to product conception, design and engineering and prototyping, as much as 90% of the costs can be essentially committed by decisions made up to this stage. Second, it is complex to the point that human skills are hard pressed to consider all issues. Even standard design problems can be highly complex in both conceptual and algorithmic terms. Integration of the design process increases the complexity of the problem by orders of magnitude. Third, the data management task for a complex product, whether mechanical or electronic, can itself be daunting. Data management includes not only the current design, but also configuration management of specific units, as well as archiving and retrieval of design data. Fourth, the increasing pressures for faster product development encourage the use of support tools which can leverage human capabilities in terms of speed as well as productivity. The reduction in the costs of design and development also creates the ability to field more products with shorter lifecycles, and to more closely match the changing needs of the marketplace. Finally, concurrent engineering by its nature often involves teams, and support tools have the effect of facilitating communication between teams, through standardization of data, terminology and concepts.

The potential benefits from concurrent engineering methods can be substantial. For example, Bogard et al. cite estimates of benefits in various areas:

- reduction of product development times of 30-70%
- improvements in quality of 200-600%
- increased white collar productivity of 20 to 110%
- reduction in engineering changes of 65-90%.

Even if only a modest fraction of these benefits are realized, the payoff from concurrent engineering can be substantial.

In the following sections, we first describe a broad conceptual framework for concurrent engineering. Next we describe the issues involved in the specific case of the design of multi-chip modules (MCM's), and present an architecture for a design support system for this activity. Finally, we discuss the development of potential modelling, optimization and analysis tools based on both AI/KBS and traditional modelling paradigms.

2. Concurrent Design and Engineering

Design processes are often represented as a hierarchy of design steps and decisions. Concurrent engineering can also be represented in these terms. Typically the design hierarchy is simply extended to cover operating issues such as manufacturability, maintainability and serviceability. In general, the design process will differ with the nature of the product area. Major differentiating factors have to do with whether the product is a component or an end use item, whether it is a consumer or industrial product, the nature of the product and process technologies (electronic, electromechanical, etc.), and the interfaces that are required. However, there are some similarities in the overall process as well. Some descriptions of concurrent engineering methodologies for electrical and electronic equipment are presented by Rosenblatt and Watson (1991), Bogard et al. (1991), and Carter and Baker (1991).

Attempts to present a general design approach have certain intrinsic limitations due to the variety of specific cases and special conditions as well as the complexity of the design process. It becomes necessary to make some sweeping simplifications while understanding why they are being made. For example, it is apparent that design is not necessarily a hierarchical sequential process. Rather, it may be highly parallel and iterative — indeed these could be crucial features of a successful design process. The process is also recursive in the sense that the entire design cycle may be repeated at different levels of a bill-of-material, in going from the finished products to major sub-assemblies to components to parts. In the interests of readability we will hereafter assume that these qualifications are understood.

A prototypical design hierarchy might contain the following levels:

- *Product Concept:* This is a statement of the market need that the product will fill and the functions it will perform. For example the product is to be a portable printer to be used with a portable or lap-top computer. This is the appropriate level to set cost and volume targets, since they will affect the design choices to be made.
- *Functional specification:* The specification of functionality should include the proposed use of the product (or component), interfaces to other products and systems, physical configuration and features, etc. To take the example of the printer, the product specification may include the size, weight, speed, footprint, printing resolution (dpi), compatibility, color, fonts, internal processing capability and use pattern (portable, laptop). There may be special design features that distinguish the product, such as say a touch screen type of control or user interface, or self configuration.
- *Technology selection:* This refers to the major choices available for the product technology. For example, in the case of a printer, the technology

selected for printing could be some version of thermal, dot matrix, ink jet, laser and so on. In addition, there may be several auxiliary or secondary technologies involved in areas like paper handling, information display, control panels, and power supplies.

- *Physical design:* This level is the core of this design hierarchy, and is the most dependent on the particular application. It involves substantial parallelism, iteration and recursion in designing the product as a collection of subsystems and components. For example, cost estimates at part and component levels become inputs to estimating total costs of the assembled product. Similarly, reliability of components and interfaces determines the reliability of the product.

In subsequent sections, we will largely address the design process at the part component and subassembly levels. However, we would make one important observation on the physical design of the complete product. One of the key problems in product design is known to be the functional separation between engineering design (which concentrates on functional performance) and manufacturability (which refers to the consequent effect of design on the total cost of manufacturing). One practical technique for integrating these aspects is to ensure that the production bill-of-materials is constructed early in the design process. This bill describes product structure in a manner that is consistent with the way the product will be manufactured, and is essential in evaluating the costs of manufacturing. Normal practice is often to have engineering bills structured quite differently from production bills, often with a simple "bucket of parts" structure.

3. Multi-chip Module (MCM) Technologies

Multi-chip modules (MCM's) represent a physical level of integration in the design of digital electronic equipment. An MCM is a set of chips packaged together so as to maximize the performance of the overall design. The reason for this level of packaging between board and chip arises from a couple of factors. First, too much specificity at the level of chip design can lead to low volume design. Furthermore, there are limits to the complexity of function that can cost effectively be achieved on a chip. Specificity, combination or generalization of functions can be achieved at the level of chip sets instead. However, integration of chips into chip sets at the level of a printed circuit board can be inefficient in terms of circuit density and the lengths of network connections. The MCM is thus a way of achieving high circuit density and integrated functional design at an acceptable cost.

There are several different kinds of MCM technologies available from corporations and research laboratories. Table 1 presents seven examples, that vary in cost and performance. Several are proprietary technologies.

Technology	Example	Status	Cost	Performance
MCM-HTCC	IBM-TCM	Mature	High	Low to Medium
MCM-LTCC	IBM	Mature	Medium	Medium to High
MCM-L	Many PWB	Mature	Low	Low to Medium
MCM-SionSi	AT&T/nChip	Mature	Medium	High
MCM-PIonSi	Hughes/APS	Mature	Medium	Medium to High
MCM-Overlay	GE/TI/ISI	Devel.	Medium	High
MCM-QTAI	MCC	Devel.	Medium	Medium to High

Table 1: Representative MCM Technologies

The MCM-HTCC is a high temperature co-fired ceramic technology that employs multiple layers of ceramic tape with vias punched and filled. The interconnect pattern is formed in a refractory metal (tungsten). The layers are aligned and fired in a high temperature oven. IBM uses this technology in their Enterprise 9000 Computer and uses the thermal condition module (TCM) packaging technology. One problem with this technology is that interconnect lines need a minimum width of 200 microns and shrinkage when fired exceeds 15%.

The MCM-LTCC is similar to the MCM-HTCC with two exceptions. Interconnections can be made in gold or silver and a low temperature (600 to 800 C) oven can be used. Line widths can be 50 microns and shrinkage when fired is < 0.2%. High frequency (1GHz) digital circuits have been fabricated using this technology.

The MCM-L (laminated) technology is widely used in Japan in the manufacture of consumer goods. Its low cost and availability from printed circuit board manufacturers makes it an ideal candidate for university/research lab prototyping activity and manufacturing. This technology is useful for medium performance (< 150 MHz) provided crosstalk between conductors can be reduced.

The MCM-SionSi is a medium cost technology that treats MCM fabrication as a giant chip rather than a miniaturized printed wire board. It uses Aluminum for interconnect wires and uses silicon dioxide or silicon nitride as dielectric. Fine lines (20 to 40 micron width) and low capacitance avoids the need for terminating resistors up to 150 MHz.

The MCM-PIonSi uses polyimide as dielectric on a silicon substrate with aluminum or copper for interconnect wires. Aluminum can be sputtered directly on polyimide or silicon dioxide but has a higher resistance than copper. Copper requires an adhesive coating to polyimide and also to protect it. The copper wires are normally electroplated. This is a mature technology and available from many vendors at a reasonable price.

The MCM-Overlay technology is radically different from the others. Alumina, AlN, or other substrate with thickness in the 1.5 to 2 mm is used. Holes as deep as 0.7 mm are milled in the substrate to place the dies. Dielectric sheets are placed on the surface and interconnects are formed. Polyimide or Kapton can be used as dielectric and they can be spun on. Multiple layers of interconnects are built one on top of another. This approach eliminates wire bonding of dies and provides good heat flow from die to substrate. This is an excellent technology for high performance. Digital circuits operating in the GHz range have been fabricated (Gdula et al., 1992). GE has made many MCMs using this technology and a foundry operation based on the technology is expected to be available in 1993 from Texas Instruments.

MCM-QTAI provides a quick turnaround interconnect (QTAI) by personalizing interconnections on a dedicated layer. It provides a standardized design for the power, ground, and x and y direction signal layers. The fifth interconnect layer is for personalizing to support the desired connectivity. This technology is developed by MCC. It also uses solid vias to improve heat conduction. Polyimide is used as a dielectric and copper as the metal for interconnect wires in this technology.

The last four technologies in the table are candidates for high performance MCM. These have been analyzed to determine ways of forming interconnection wires in the substrate. Five ways for forming interconnects on the MCM substrate have been identified.

- *Fixed interconnection:* This approach employs five layers of interconnection wires in the substrate: one layer for power, one layer for ground, two layers for interconnecting circuits, and one layer for connecting to the pads of dies. The AT&T-AVP process, nChip's MCM-SionSi process, and Raychem's APS process use this scheme.
- *Mask programmable interconnection:* This approach also employs five layers of interconnection wires in the substrate: one layer for power and one

layer for ground. Two interconnect layers are laid out in horizontal and vertical segments and the edges of the segments are brought out to the third layer using vias. Connections between adjacent horizontal or vertical segments, and horizontal and vertical segments are formed in the third layer using a custom mask. The connecting points to pads are also on the third layer. The MCM-QTAI process of MCC is an example of this approach.

- *Field programmable interconnect:* This is a novel approach that allows programmability to the interconnection wires either in the factory or in the field. The first four layers are similar to that employed in the mask programmable interconnection scheme. The fifth layer is used for connecting the segments on the third and fourth layers using antifuse technology and connecting to pads of dies. The antifuse technology allows one time programmability only.
- *Custom interconnection:* The number of interconnection layers available can range from 4 to 15 for some of the MCM processes. A layout system is responsible for routing each layer to achieve the desired connectivity. The interconnection patterns are specified by the design files in Gerber, Finesse, or CIF for use by the MCM foundry. The MCM-Overlay process of GE and the MCM-PIonSi process of AT&T, Raychem's APS and others are well suited for this approach.
- *Custom interconnection with passive components:* This approach is similar to custom interconnection with the addition of passive components such as bypass capacitors and termination resistors. The MCM-SionSi technology of nChip and the MCM-Overlay technology of GE are well suited for this approach.

Many techniques are available in the industry and research laboratories for attaching components to pads on the substrate of a MCM. The following techniques are the most relevant because of their availability and reduced cost.

- Wire bonding
- TAB
- Flip TAB
- Flip chip
- Demountable TAB (DTAB)

For further description of MCM technologies see Gdula et al. (1992), Roszel and Daum (1992), Frye et al. (1992), Shambrook (1991, 1989) and Lang (1991).

4. Architectures for Concurrent Engineering and MCM Design

The development of a support environment for concurrent engineering and design of MCM's requires complex information systems which must be viewed from several different structural perspectives. Figure 1 places the MCM design problem in a broader context. The overall concurrent engineering task proceeds from specification of the product, its architecture, to the design of chips and modules.

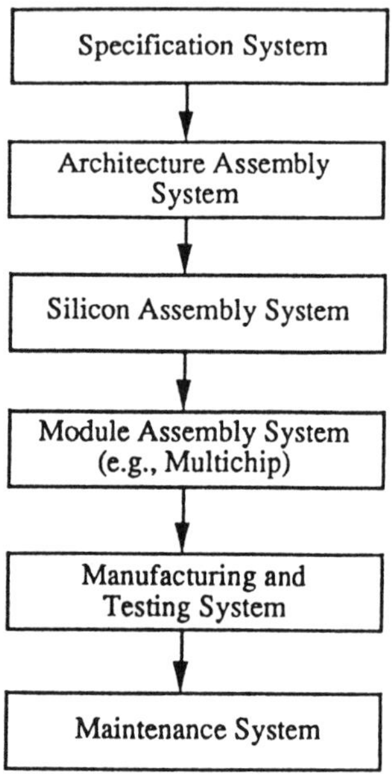

Figure 1: Subsystems of a Concurrent Engineering System

In Figure 2 we present a reference architecture for a concurrent engineering and design system for MCM's. The vertical slices represent application tools integrated via a data manager which provides both archiving and communication between tools. The horizontal slices are described briefly below:

The *procedural* layer is used to define and validate the consistency of the requirements, particularly with respect to the "user" interface. This high level evaluation leads to partitioning the overall requirements into a set of *architectural* and *functional* specifications and a corresponding set of system level blocks. These specifications are then further partitioned in terms of a *register transfer* and *structure* level based on interconnected modules. Typical building blocks at this stage are storage/memory elements, arithmetic elements and J/O elements. These building blocks are then composed from a macro cell library at a *device* and *mask* level. Note that at every stage the designer has the necessary facilities to validate the design and the requirements using dynamic simulation and other related tools.

Data Manager	Commercial Tools	
	I- Logix	Procedural Layer
	Network II.5, COMNET	Architectural Layer
	SES/design, sim, scope workbench	Functional Layer
	VHDL (MCC, Vantage) Mentor's Idea Station	Register Transfer Layer
	Mentor's Design Architect	Structural Layer
	Mentor's Analog Station Hybrid Station, MCM Station, IC Station	Device Level Layer
		Mask Level Layer
	PCB Place, Parasitic Parameters, Omnicard **Racal** **Dazix**	Packaging Layer
	EEsof, Contec	Electrical Parasitic Extraction
	Aries, PCB Thermal, Vibration PCB Fatigue, Soldering Simulator, Hybrid Ther. SDRC, Pacific Numerix	Thermal, Mechanical, Environmental Layer
	AT&T's GENTEST	Testing Layer
	TI's CAS DFM	Manufacturing Layer
		Maintenance and Repair Layer

Figure 2: Commercial CAD Tools and CE Layers with Data Management Interface

At this point in the design process the designer can explore *packaging* and placement options. Advanced IC packaging options include Slipchip, SlipTAB and TAB. This results in a baseline design that must then be analyzed in terms of *electrical parasitic*, *thermal*, *mechanical* and other *environmental* issues. Electrical parasitic effects include high frequency transmission line delays; thermal issues include heat conduction from junction hot spots; and mechanical issues include vibration effects on packaging. As a result of this evaluation, some of the ad hoc design assumptions made at an earlier stage of the design process may well have to be revised. In general, the design framework must support this type of design revision in a feedback or feedforward informational flow.

Having completed the design up to this stage, the designer must now consider insertion of test procedures, test points and related manufacturing issues pertaining to module level partitioning/fabrication/testing. Other important considerations include maintenance and repair (e.g., LRU - line replacement unit), testing procedures, and spares strategy.

In summary the entire CAD/CE process must be able to support a flexible data management structure that is comprehensive, extensible and robust.

Commercial tools that are currently available are shown in the appropriate layer. While there are many such tools, which provide various capabilities, an integrated design application is lacking. Furthermore, there is at present not integrative data management tool available, though such tools are being proposed. It seems likely that in the future, the development of specialized design, support and evaluation tools will continue to take place in a decentralized manner. An integrative data management environment that can accommodate such developments is an essential component of a CE system.

5. MCM Design Methodology and Tools

There are two main approaches to designing an MCM system. The first approach treats MCM design as an extension of PCB design containing surface mounted parts with a new set of constraints. This approach is taken by many of the commercial CAD vendors. The MCM tools provided by Mentor, Cadence, and Intergraph fit this style. The Honeywell Finesse set of tools for MCM-Overlay process of GE also employs this approach.

The second approach treats MCM design as a VLSI design with the MCM design and early analysis at the top level and chip design at lower levels. The second approach is selected for this project. Very few commercial tools take this approach at this point although they may head in this direction a few years from now. The set of tools needed for MCM design and implementation of a digital system using this approach is shown in Figure 3. Many factors have to be considered in designing tools for high performance systems. Since different sets of tools are needed for different levels of abstractions a cohesive approach is needed in designing these tools to achieve high performance in systems designs. A set of performance goals and constraints has been formulated for use by tool developers. The goals and constraints reflect MCM technologies, MCM assembly robots, manufacturing techniques, cooling techniques, and substrate material. The key objectives are:

G_1: Reduce interconnect propagation delay. This delay is from the output pad of a die to the input pad of another die or primary output. (Function of wire length, frequency and capacitive coupling with other wires.)

G_2: Reduce total area on the MCM. (Function of area used for attaching and connecting dies to substrate and wire routing among dies.)

G_3: Reduce total power consumed by the MCM.

G_4: Reduce the total wire length in the MCM. (Function of area and chip placement.)

G_5: Reduce the total cost of MCM. (Function of area, layers, vias, technology.)

Note that these objectives are interdependent.

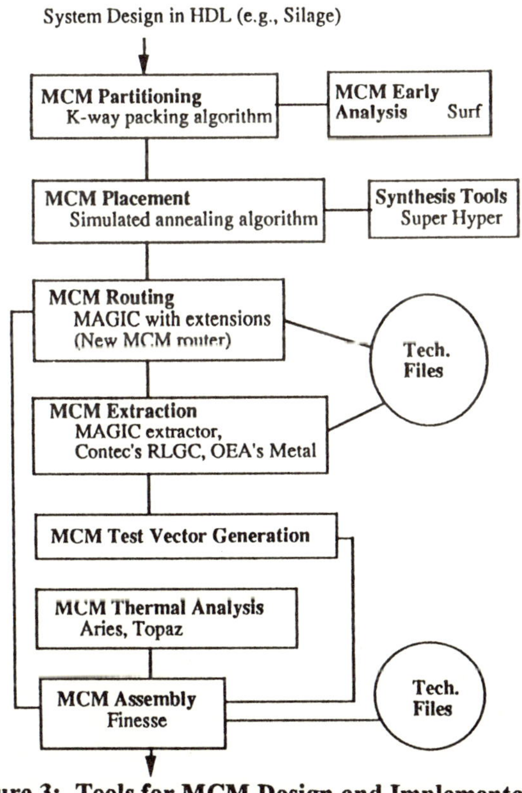

Figure 3: Tools for MCM Design and Implementation

The objective function to be minimized could be defined as a linear combination of measures for these five goals, G_1 to G_5. For example,

$$OBJF = c_1 * G_1 + c_2 * G_2 + c_3 * G_3 + c_4 * G_4 + c_5 * G_5$$

The coefficients c_1 to c_5 indicate the weights assigned to each of these measures. Minimizing OBJF is a global optimization problem, while reducing G_1 to G_5 can be treated as local optimization problems. It may not be possible to achieve simultaneously optimal values for G_1 to G_5 and OBJF. Each design tool has to address this problem based on the constraints imposed at the appropriate level of abstraction.

The constraints to be satisfied are the following:

- Packaging constraints for components (die interconnects, and heat sinks).
- Required minimum distance between parts (clearances needed by pick and place robots and for inspection and testing).
- Number of interconnection layers that can be supported by the chosen MCM technology.
- Assembly techniques and their requirements (solder reflow, inprocess inspection and test).
- Required heat dissipation from die (allowable junction temperature).
- Restrictions on cooling techniques (e.g., only forced air cooling).
- Cost constraints and vendor constraints (e.g., use 7404 dies from TI).

MCM Partitioning. Partitioning is an important step for the hierarchical design of complex digital systems. Large designs cannot be economically implemented as a single chip because of yield problems. The design is partitioned along functional, bit-slice, random, recursive min-cut, or clustering basis into multiple chips, placed on an MCM, and routed. System partitioning under timing and capacity (area or number of pins per chip) constraints is needed to produce a quality system. The input to the partitioner is a "netlist" or a specification in HDL. The partitioning problem is stated as: Given an MCM with the capacity for K chips and a timing constraint T, to map the system expressed by a graph G with N nodes and E edges to the K chips so that the number of wires crossing the chips is reduced and the timing constraint T is satisfied.

One approach to solve this problem is to generate a new system graph where the nodes are clustered without violating the timing constraint. The modified graph is then mapped to the K chips using a K-way packing algorithm such as those proposed by Kernighan and Lin (1970), Wei and Cheng (1989), and Shih, Kuh and Tsay (1992) to reduce wire crossings while maintaining timing constraints. Other constraints such as heat dissipation and

I/O pin constraint can also be added to the approach at the expense of complicating the algorithm.

MCM Placement. Floorplanning and the actual placement of parts is a key step in MCM design. A poor floorplan can affect performance by increasing interconnect delays on critical nets and thermal hot zones. Many tools are available for floorplanning and placement of cells in a VLSI design. They allow interactive and automatic floorplanning and placement, global routing, and power and ground routing. Although they can be used for MCM placement they have some limitations in routing. Research tools are available for placement and floorplanning of VLSI/ULSI circuits designed in the macrocell approach. They can be used for MCM where the individual chips on the MCM are treated as macrocells. These methods use, for example, a simulated annealing algorithm and allow constraints such as critical nets and maximum net lengths to be specified.

MCM Routing. Routing dies on an MCM is a challenging activity. The first challenge comes from the number of layers (> 6). The next challenge comes in the need to accommodate many types of vias (stacked, buried, staggered, and filled). High electrical performance requires interconnection lines to be designed as transmission lines. Long parallel lines can cause cross talk and the router must detect and avoid such lines. Estimates of lengths of lines are needed before determining their widths. Distance between lines must also be determined to reduce cross talk. Another challenge is in redistributing empty space available locally so that lines are evenly spaced. This conflicts with the goal of reducing spacing between lines and must be carefully done. Some MCM technologies allow thermal vias to be inserted to improve heat dissipation. The router has to be smart enough to accommodate thermal vias. Another challenge is to reduce parasitic electrical effects of interconnection lines.

Global routers have been used in the 1960's (Lee, 1961) to design boards for such as IBM 360/370, CDC 6600, and Cray 1-S in the 1970's. The high performance systems have some requirements for global routing. Interconnect lines should be free of right angle turns. Clocks should be routed on a separate layer. Clock signals must be distributed as differential pairs to keep signal integrity. Interconnection lines representing critical nets must be routed next to a ground layer. These critical lines must also be surrounded by guard traces. It is desirable to route these critical lines on the top metal layer. Vias in interconnection lines cause reflections and they should be avoided in clock lines. The power and ground layers should use wide lines. There are many other requirements and these can be found in the literature. Although signals such as clock account for less than 5% of the interconnection lines in a system, using a separate layer is one of the techniques that can pay off handsomely in the long run.

In VLSI design, global routers such as the LocusRoute (Rose, 1988, 1990) provides efficient routing for standard cell based designs. Extending these routers to MCM may be feasible. For example, Magic (Ousterhout et al., 1985) is an interactive system for creating and modifying VLSI circuit layouts. It has routing tools that allow global interconnection in circuits. It uses Manhattan geometry (horizontal and vertical lines, no 45 degree lines). The use of Manhattan geometry costs 5 to 10% in density but reduces design time. Magic uses a grid based channel router. It knows how to avoid obstacles by routing in a layer under, over, or around and the user can control the quality of routing. However, to support MCM routing the basic routing algorithm of Magic would have to be changed.

A general purpose router is needed for efficiently routing large MCM design. This router needs to solve the following MCM routing problem: Given a net list <N, E>, where N is a set of nodes represented by a 6-tuple, (chip_instance_name, location represented by x and y coordinates, pin number, side_of_chip, pin_relative_position), and E is a set of connections represented by the 3-tuple, (source_node, destination_node, constraint_list); implement the connections using layers metal1, metal2, ..., metaln such that the total wire length is reduced, number of vias is reduced, and the constraint list is satisfied for each of the connections. The pads for the chips are assumed to be in layer metal0. The objective function OBJF discussed earlier is to be used by the router in doing its work.

Routing algorithms addressing the above problem are in various stages of development in industry and universities. Since interconnection lines in MCMs can have variable width to achieve high speed, gridless routers are desired. The router must estimate the length of lines before determining their widths. The router should be able to detect long parallel lines in the same layer or on adjacent layers and avoid them to reduce cross talk. The spacing between long parallel lines in the same layer may have to be increased to reduce cross talk. If thermal vias are allowed in the MCM technology the router should be able to route around the vias. The same applies to critical net routing. It is necessary to determine the topology of each net starting with the critical nets and assign them to layers. The less important nets are then assigned to layers. A topological routing is needed to perform global refinement of die positions and via location. In doing this the interconnect lines are treated as constraints. The geometrical routing is to be carried out at the end to complete the routing.

The global routing scheme proposed by Dai (1991) for MCM appears to address the above requirements. The notion of a sketch to represent the topological routing of one layer is introduced. A sketch contains rigid objects, called features, and flexible wires. A routing topology may be represented by many different sketches and a canonical form, called rubber-band sketch is introduced. A scheme for producing legal routings of rubber-band sketches is discussed.

MCM Extraction. The purpose of MCM extraction is to obtain the interconnect resistance, capacitance, lead inductance, and parasitic capacitance. These parameters are used to estimate interconnect delays. An example is the UC Berkeley Magic extractor for VLSI technologies and the MCM-PIonSi technology of AT&T. The Magic system extracts a lumped resistance for each node, rather than a point to point resistance between each pair of entities connected to that node. This results in worst case resistance values to be used for point to point resistances. Three kinds of internode coupling capacitors are extracted. They are overlap capacitance (parallel plate capacitance between overlapping material), sidewall capacitance (parallel plate capacitance between the vertical edges of two pieces of the same material), and sidewall overlap capacitance (orthogonal plate capacitance between the vertical edge of one piece of material and the horizontal surface of another material that overlaps the first edge).

Interconnect Analysis. High performance digital systems are heavily dependent on their interconnection lines for performance. The interconnect lines can be treated as lossless transmission lines for frequencies up to 5 GHz. Another way is to determine the wavelength (speed of light/frequency of circuit operation) and compare it to the length of interconnect lines. If the length of lines is greater than wavelength then the transmission line model is to be employed. Otherwise, analog circuit simulators such as Spice can be used for circuit analysis. Many parameter extractors and interconnect analyzers are available from commercial EDA vendors. Some examples are HP's structure simulator (1990), EEsof's Xtract and EMsim (1992), TMA's RAPHAEL interconnect analysis system (1991, 1992), OEA's Metal 2D and GDSII system (1992), and Contec's RLGC extractor (1992) and signal integrity analysis software.

MCM Test Vector Generation, Thermal Analysis, and Assembly. The testing of MCMs and automated test generation are still research issues. Thermal analysis is an especially difficult task. There are some good commercial tools available from Pacific Numerix, Aries and others. These tools use a finite element method to model 3-D heat transfer in MCMs. The integration with MCM assembly analysis remains an open issue.

REFERENCES

Aptix (1992), "Bringing Programmability to the System Level", Aptix Internal News Release, 225 Charcot Avenue, San Jose, CA, (May).

Bogard, T., B. Hawisczak and T. Monro (1991), "Concurrent Engineering Environments: Texas Instruments Implements Design for Manufacture", Printed Circuit Design, 30-43.

Brodersen, R. W. and et al. (1991), "Lager IV Reference Manual", UCB Software Release, University of California, Berkeley, CA, (June).

Burns, J., A. Casotto and et al. (1987), "MOSAICO: An Integrated Macro-cell Layout System" UBC Memo, UCB/ERL M8717, University of California, Berkeley, CA, (January).

Bursky, D. (1992), "Programmable Interconnection Matrix in Silicon Speeds System Design", Electronic Design, (January), 31-32.

Carter, D. E. and B. S. Baker (1991), Concurrent Engineering: The Product Development Environment for the 1990s, Addison-Wesley Publishing Company.

Contec (1992), "Tool Targets Transmission Line Effects in High Speed Designs", Computer Design, (July), 109-110.

Dai, W. (1991), "Performance Driven Layout of Thin-film Substrates for Multichip Modules", Proceedings of the 1991 IEEE Multi-Chip Module Conference, MCMC-91, Santa Cruz, CA, (March 28-29), 114-121.

EEsof (1992), "EEsof's High-Frequency CAE Software Puts it All Together", EEsof Brochure, 5601 Lindero Canyon Road, Westlake Village, CA.

Frye, R. C., K. L. Tai, M. Y. Lau and A. W. C. Lin (1992), "Silicon-on-Silicon MCMs with Integrated Passive Components", Proceedings of the 1992 IEEE Multi-Chip Module Conference, *MCMC-92*, Santa Cruz, CA, (March 18-20), 155-158.

Gilbert, B. K. and G. W. Pan (1992), "Packaging of a Ga As Signal Process on Multichip Modules", IEEE Transactions on Components, Hybrids, and Manufacturing Technology, Vol. 15, No. 1, 15-28.

Gdula, M., W. P. Kornrumpf and B. K. Gilbert (1992), "An Overlay Interconnect Technology for 1GHz and Above MCMs", Proceedings of the 1992 IEEE Multi-Chip Module Conference, MCMC-92, Santa Cruz, CA, (March 18-20), 171-174.

Hagge, J. K. (1992), "State-of-the-Art Multichip Module for Avionics", IEEE Transactions on Components, Hybrids, and Manufacturing Technology, Vol. 5, No. 1, 29-42.

HP (1990), "High-Frequency Structure Simulator", Hewlett Packard Company Brochure, Palo Alto, CA.

Kernighan, B. W. and S. Lin (1970), "An Efficient Heuristic Procedure for Partitioning Graphs", Bell System Technical Journal, Vol. 49, No. 2, 291-307.

Lang, D. (1991), "Cost Effectiveness of nChip's MCM Technology", Proceedings of the 1991 IEEE Multi-Chip Module Conference, MCMC-91, Santa Cruz, CA, (March 28-29), 16-23.

Lee, C. Y. (1961), "An Algorithm for Path Connections and its Applications", IRE Transactions on Electronic Computers, Vol. EC-10, 346-365.

OEA (1992), "Metal 2D and 3D Interconnect Simulator", OEA International Inc., 3235 Kifer Road, Santa Clara, CA.

Ousterhout, J. K., W. S. Scott, R. N. Mayo and G. Hamachi (1985), "1986 VLSI Tools: Still More Works by the Original Artists", UCB Report, UCB/CSD 86/272, University of California, Berkeley, CA, (December).

Rabaey, J. M. and et al. (1991), "Fast Prototyping of Datapath-Intensive Architectures", IEEE Design & Test of Computers, (June).

Rose, J. (1990), "Parallel Global Routing for Standard Cells", IEEE Transactions on Computer-Aided Design, Vol. 9, No, 10, 1085-1095.

Rose, J. (1988), "LocusRoute: A Parallel Global Router for Standard Cells", Proceedings of the 25th Design Automation Conference, (June), 189-195.

Rosenblatt, A. and G. F. Watson (1991), "Concurrent Engineering", IEEE Spectrum, 22-37.

Roszel, L. E. and W. Daum (1992), "MCM Prototyping Using Overlay Interconnect Process", Proceedings of the 1992 IEEE Multi-Chip Module Conference, MCMC-92, Santa Cruz, CA, (March 18-20), 36-39.

Shambrook, K. P. (1991), "Overview of Multichip Module Technologies", Proceedings of the 1991 IEEE Multi-Chip Module Conference, MCMC-91, Santa Cruz, CA, (March 28-29), 1-9.

Shambrook, K. P. (1989), "High Density Multichip Interconnect", IEEE 39th Electronics Components Conference, Houston, Texas, (May 22-24).

Shih, M., E. S. Kuh and R. Tsay (1992), "System Partitioning for Multi-Chip Modules under Timing and Capacity Constraints", Proceedings of the 1992 IEEE Multi-Chip Module Conference, MCMC-92, Santa Cruz, CA, (March 18-20), 123-126.

Srini, V. P., J. V. Tam, T. M. Nguyen, Y. N. Patt, A. M. Despain, M. Moll and D. Ellsworth (1987), "A CMOS Chip for Prolog", Proceedings of the International Conference on Computer Design, Rye Town, NY, (October), 605-610.

Tai, K. L. (1991), "Si-on-Si MCM Technology and the Initiation of A University MCM Program", Proceedings of the 1991 IEEE Multi-Chip Module Conference, MCMC-91, Santa Cruz, CA, (March 28-29), 10-13.

TMA (1992), "RAPHAEL Interconnect Analysis", Technology Modeling Associates Inc. Brochure, Palo Alto, CA.

TMA (1991), "Technology CAD", Technology Modeling Associates Inc. Newsletter, Palo Alto, CA.

Watari, T. and H. Murano (1985), "Packaging Technology for the NEC SX Supercomputer", 35th Electronics Components Conference, Washington, DC, 20-22.

Wei, Y. C. and C. K. Cheng (1989), "Towards Efficient Hierarchical Design by Ratio Cut Partitioning", Proceedings of IEEE International Conference on Computer-Aided Design, Santa Clara, 298-301.

Winner, R. I. and et al. (1988), Institute of Defense Analysis Technical Report, IDA-R-338, NTIS Access No. AD-A203 615/0/XAB.

9

BENEFITS AND LIMITATIONS OF STRUCTURED METHODOLOGIES IN PRODUCT DESIGN

Philip Barkan, Professor
Stanford University

Martin Hinckley, Graduate Student[1]
On Leave, Sandia National Laboratories

Abstract: The remarkable effectiveness of Japanese practices has led to a growing interest in the application of structured methodologies to the processes of product design and manufacture. In this paper we will critically examine some of the unexpected benefits and pitfalls which have been experienced by some major U.S. companies in the application of these methods to current products. While unexpected benefits have been achieved, it is also shown that rigid application of oversimplified rules has been seriously counterproductive.

From an awareness of these experiences we have been led to an ongoing inquiry into the true nature of design simplicity. By treating the Pareto distribution in probabilistic terms and as a discrete embodiment of the Riemann zeta function, we have been able to quantify some aspects of design simplicity and thereby to account for a recently reported, empirical relationship between product design attributes and observed rates of defects in product assembly. Quantitative insights derived from this study identify the most effective design/manufacturing strategies to reduce defects as a function of the currently existing magnitude and source of defects.

---

[1] Work supported by the United States Department of Energy under contract DE-AC04-76DP00789.

1. Introduction

Structured methodologies are widely associated with successful product development. Such structured methodologies as Functional Analysis[2], Quality Functional Deployment[3], Robust Design[4], and Design for Assembly[5] have contributed to the rapid design of highly attractive products that have been well suited for the target market while combining low cost and high quality. As a consequence there has been a justifiable kindling of interest in such methodologies by U.S. corporations as they strive to enhance their competitiveness.

In this paper we will focus on two aspects of one of the most popular design methodologies, Design for Assembly (DFA). In Section 2, we will first examine some cases of shortfall in important applications of DFA - cases where rigid adherence to the structured rules of the methodology have proven counter-productive; we will then seek insights into the lessons which can be learned.

In Section 3 of the paper we will consider a new relationship between DFA criteria and defect rates in assembly. Starting from some observations reported by a major manufacturer we will describe the development of a new quality improvement tool that is useful in the early design stage for predicting the probable defect rates in assembly and for devising an effective strategy of quality improvement. This new methodology should prove helpful in deciding on the most effective allocation of effort to reduce product defects.

2. Problems arising from blind application of DFA design rules

Structured methodologies and design rules may provoke a tendency to oversimplify and may become a substitute for in-depth, global thinking. Whenever satisfying rules or complying with a methodology becomes the end rather than the means, objective inquiry and critical thought may be cut off, posing a danger to the product development effort.

In other cases, the methodologies may be applied superficially, leading to serious oversights and unanticipated difficulties resulting from overlooking essential considerations or an inadequate grasp of the design rules. Limited

---

[2] G. Pahl and W. Beitz, Engineering Design, (London: The Design Council, 1988), pp. 66-82.

[3] Bob King, Better Designs in Half the Time, 3rd ed., (Methuen: Goal QPC, 1989).

[4] Phillip J. Ross, Taguchi Techniques for Quality Engineering., (McGraw-Hill, 1988).

[5] G. Boothroyd and P. Dewhurst, Product Design for Assembly, (Kingston, Boothroyd & Dewhurst, Inc., 1985).

exposure to the available methodologies may result in the selection of a technique that is less than optimum for the intended application. The need is to understand the risks and to learn how to use properly the many valid strengths of structured methodologies.

In the spirit of continuous improvement, and with the benefit of hindsight, we will critically examine some specific cases where problems have arisen from the application of DFA and seek useful generalizations.

2.1. Development of a US manufactured Dot-Matrix Printer. The original development of a cost-effective, domestically manufactured dot-matrix-printer for the personal computer market in the early 1980's was widely regarded as a great success story, and as an early example of effective concurrent engineering. However, there were some significant shortfalls in the project which can offer important learning experiences and illustrate the hazards of rigid adherence to design rules.

In this instance, the product development team was severely constrained to adhere to several arbitrary, but rigidly enforced, DFA design rules. These rules were formulated with the intent of developing in a minimum time period a cost-competitive printer to be manufactured in a dedicated, highly automated plant. Compliance with each rule produced benefits and penalties. The impact of rigid application of a single rule drawn from the DFA structured methodology is illustrative of the types of problems encountered.

Among other requirements the design team was instructed to: Avoid flexible elements that are difficult to handle, such as coil springs and belts. To conform to this requirement, a molded part eliminated the need for a flexible coil spring by incorporating a very thin cantilever beam as an integral element of the complex part. The design was extremely successful in conforming to the rule; however, this success was clouded by several fundamental issues which revealed the risks of such a narrow focus.

The cantilever beam incorporated in the plastic part as a replacement for the spring had to be extremely thin in order to meet functional requirements, and thus was necessarily fragile. This fragile component was highly stressed and subject to long term creep resulting in loss of function. Its extreme fragility increased the likelihood of failure. Failure of this single element required replacement of the entire, complex and costly side frame. Thus a reliability problem and extremely costly repair were introduced in the zeal to comply with arbitrary rules that eliminated anticipated handling problems with a relatively robust simple spring costing a few cents.

To meet the multiple functions demanded of this single part, extremely tight tolerances, far closer than routinely attainable in molded plastic parts, were required. The solution required complex molds and heavily filled plastic compounds which led to extremely costly development and manufacture with increased maintenance problems. Mold life was severely reduced due to excessive wear from the heavily filled molding compounds.

In hindsight it is evident that the design rules were too rigidly applied. It became a point of prestige within the team not to be the first to violate a rule, creating a climate which did not allow the rules to be questioned or resisted. The rules became the ends unto themselves. By employing structured methods as goals rather than means, the essential needs and purposes were overlooked. Too many innovations were simultaneously embraced, and long term problems became inevitable.

2.2. Part Count Problems in the Design of an Ink Jet Printer. A second example from another major corporation again illustrates the fallacy of rigidly applying the design rule to minimize the part count. In this case, the following insights were gained in a post-mortem evaluation conducted with the design team:

> "In particular they felt that they had placed too much emphasis on minimizing the number of parts to the point where too much functionality was packed into the plastic chassis piece. It would have been less expensive to have more parts, hire more direct labor to assemble those parts, and reduce the amount of engineering time involved in developing the complex part. Tooling is wearing more rapidly than expected. In retrospect they believe that three parts would have been better than one, relieving some of the tooling tolerance problems they are now suffering."[6] (Underlining added)

This example demonstrates again the problems which can occur when blind adherence to the rule becomes a higher priority than sound product design. In many cases it is possible to combine parts to reduce part count, but this is not always a wise decision as the following example aptly illustrates.

2.3. Comparison of Instrument Panels for Two Competing Truck Designs. Similar insights were gained from a recent independent study of design features found in instrument panels of two competing light trucks - one domestic and

---

[6] Sara Beckman, Draft of HP Chapter for Manufacturing Visions Group, July 16, 1991.

the other imported. Comparing the two products, researchers arrived at the following conclusions:

> The import had more parts but probably simpler mold designs. A rough qualitative assessment indicates that their tooling costs are significantly lower than that of the domestic product. Interestingly, the import panel contained less snaps than the domestic panel. While the import panel contained more parts it was assembled in less time.[7]

These observations boldly contradict some fundamental principles espoused by DFA. The import product had more screws and roughly 20% more parts than the domestic product, but could be assembled in less time. The study revealed that the two products had roughly the same number of assembly operations, but the import operations were generally simpler and less time consuming. In several instances, the larger part count allowed significantly simpler part fabrication and simpler assembly operations

2.4. Lessons Learned. From these few case studies we can see how DFA rules, and particularly pursuit of minimum part count, can lead to serious improper emphasis in product design. A global perspective must prevail in evaluating the consequences of design decisions. There will always remain a tension between time to market, product flexibility, long term cost, quality and performance, and the need for product innovation, etc. These are complex, difficult issues which must ultimately be reconciled in evaluating the impact of adhering to design rules.

To achieve this goal, "*global simplification*" must be regarded as a higher criterion than local optimization. Design "rules" and methodologies should be viewed as guidelines with allowance for flexibility. Management can play a critical roll in promoting an attitude which accommodates deviations from the rules when appropriate. As the previous examples have illustrated, parts should only be combined if the result is *global simplification*.

To this point, we have focused on the down side and some severe risks of oversight. We will now examine a quite different situation where unexpected benefits can be derived from a sound application of the DFA methodology.

---

[7] Steve Blum, Ade Mabogunje, and Brian Reilly, Design for Manufacturability Study, Stanford University, ME217B Class Final Report, June 20, 1991, pp. 3 and 34.

3. Design Assembly Efficiency and Assembly Defect Rates

Practitioners tend to focus on part count as the measure of design effectiveness.[8,9] But, as has been amply shown in the preceding sections, part count is an inadequate and potentially dangerous focus for design. If part count is not a valid criteria, what is?

Assembly efficiency is a parameter introduced by Boothroyd and Dewhurst[10] in their DFA structured methodology. Assembly efficiency compares computed assembly times to an ideal but arbitrary standard. This relationship is expressed as follows:

$$EM = \frac{t_{ideal} \cdot NM}{TM} = \frac{3 \text{ seconds} \times NM}{TM} \qquad (1)$$

Where, t_{ideal} = B&D ideal assembly time per operation (3 sec)
EM = the manual assembly efficiency
NM = the theoretical minimum number of parts
TM = the total manual assembly time in seconds

Assembly efficiency can be interpreted as a measure of the potential to achieve further reduction in assembly time by redesign. In general, this parameter has been little appreciated and its significance is frequently ignored in practice.

Recently, a remarkable relationship was observed between the defect rate in the factory assembly of several mass produced electro-mechanical products and the assembly efficiency. The relationship reported by Motorola is shown in Figure 1,[11] and reveals a clear correspondence between the assembly efficiency rating of a given product design and defect rates encountered in production assembly. Motorola did not offer any explanation for this rather remarkable relationship.

Our interest in this data was peaked by the possibility that such a relationship could provide a basis for a general, quantitative predictive tool. Such a tool could be used in the earliest stages of concept development to estimate probable assembly defect rates, allowing comparison of competing

---

[8] Douglas Dactz, "The Effect of Product Design on Product Quality and Product Cost", Quality Progress, June 1987, pp.63-67.

[9] James G. Bralla, Handbook of Product Design for Manufacturing, (New York: McGraw-Hill, 1986), pp. 63-67.

[10] Boothroyd and Dewhurst, p 2-3.

[11] Bill Brannan, "Six Sigma Quality and DFA-DFMA Case Study," Boothroyd and Dewhurst DFM Insight, 2 (Winter 1991), pp. 1-3.

concepts or stimulating redesign at the time when it is easiest to make design changes.

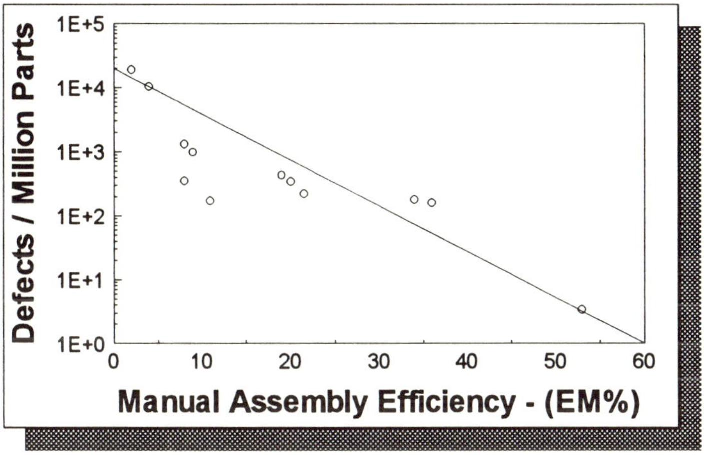

Figure 1: Observed Defects per Million Parts versus the Manual Assembly Efficiency published by Motorola.[12] The line drawn by Motorola is not known to be based on a curve fit.

3.1. Development of a theoretical basis for Probable Defect Rate Prediction. It is beyond the scope of this paper to describe the development of the theoretical basis relating the probable defect rate per assembly operation to assembly efficiency. We can however briefly outline the underlying ideas and focus on an interpretation of the results. Based on the Motorola data, we anticipated that some of the product defects could be related to the duration of assembly operations. Subsequent theoretical development was based on the postulate that in any design, the set of manual assembly times per operation follow a Pareto distribution.

Plotting process variables in a Pareto Chart[13,14] typically shows that a small fraction of the factors account for most of the variation. A Pareto chart of

---

[12] Ibid.

[13] Kaoru Ishikawa, Guide to Quality Control, (White Plains: Asian Productivity Organization, 1982), pp. 42-49.

[14] J. M. Juran and Frank M. Gryna, Juran's Quality Control Handbook, (New York: McGraw-Hill, 1988).

assembly times per operation for a typical DFA evaluation is shown in Figure 2.

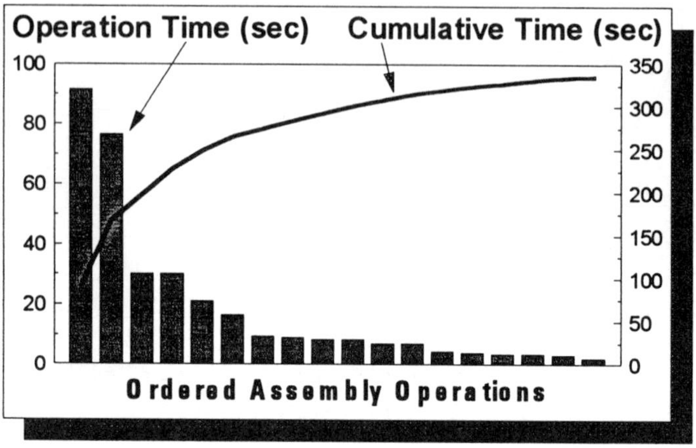

Figure 2. Pareto Chart for a series of assembly operations. Data is from a study of Thermal Conductivity Test Equipment. Individual assembly operation times represented by bars are plotted with the scale on the left. The scale on the right is for the cumulative assembly time shown as a line.

In general, as the number of operations in an assembly increases, there is a higher probability that some of the assembly steps will be more complex and will require longer assembly times than the ideal. The Pareto chart shown in Figure 2, which is a fairly typical example, demonstrates that the shortest assembly times are the most likely to occur.

3.2. The Riemann Zeta Function and the Pareto Principle. The Pareto chart acquired its name from an Italian economist who demonstrated that a discrete Riemann zeta function could be used to describe the distribution of incomes in a given country.[15,16,17] G. K. Zipf popularized its use by demonstrating that these distributions have broad application.

---

[15] Sheldon Ross, A First Course in Probability, 3 ed., (New York, Macmillan, 1988), p. 142.

[16] I. S. Gradshteyn and I. M. Ryzhik, Table of Integrals, Series, and Products, (London: Academic Press, 1980), pp. 1072-1074.

Statistical methods have confirmed that assembly operation times follow a Riemann zeta distribution described by two adjustable coefficients while rejecting at highly significant levels the possibility that assembly operation times are normally distributed. This demonstrates that the Pareto distribution may be used as a *predictive* tool in this application.

When an appropriate probability distribution has been identified, there are standard methods for estimating the outcome of a series of random "trials" or selections from the distribution.[18] These techniques can be used to define a probabilistic relationship between assembly time and the number of parts or assembly operations.

The Pareto relationship has been used to bound the assembly efficiency as a function of the number of assembly operations as illustrated in Figure 3. This relationship provides a critical missing link that relates a measure of product simplicity to changes in the part and operation count. Without this relationship, it is impossible to accurately compare concepts which differ in the number of parts, operations and/or estimated assembly times.

As shown in Figure 3, over a wide range the assembly efficiency can be determined within relatively narrow limits. This provides a powerful analytical tool useful during concept development. By simply determining the number of operations and theoretical minimum number of parts in each product concept, the limits of assembly efficiency can be determined with a minimum amount of analysis and without any dependence on a database.

Note that there is a general improvement in assembly efficiency when the operation count ratio is reduced. To achieve efficient assembly, the operation count ratio must be less than two. By contrast, when the part count is minimized, assembly efficiency may still be very low. A study of more than 150 assemblies revealed that many designs with the theoretical minimum number of parts frequently had very low assembly efficiency.

This demonstrates that minimizing part count frequently does not lead to reasonable assembly efficiency, and in some cases there is a probability that it may lead to degradation of the assembly process. Thus, the answer to the question of a more valid design criterion than part count is provided by Figure 3. *Minimizing and simplifying assembly operations* will tend to reduce part count, simplify part interfaces, and avoid inefficient assemblies. This results in a more global simplification than minimizing part count can achieve.

---

[17] CRC Handbook of Mathematics, 6 ed., (Boca Raton: CRC Inc.), p. 78.

[18] Sheldon Ross, A First Course in Probability, 3 ed., (New York, Macmillan, 1988), pp. 108-142.

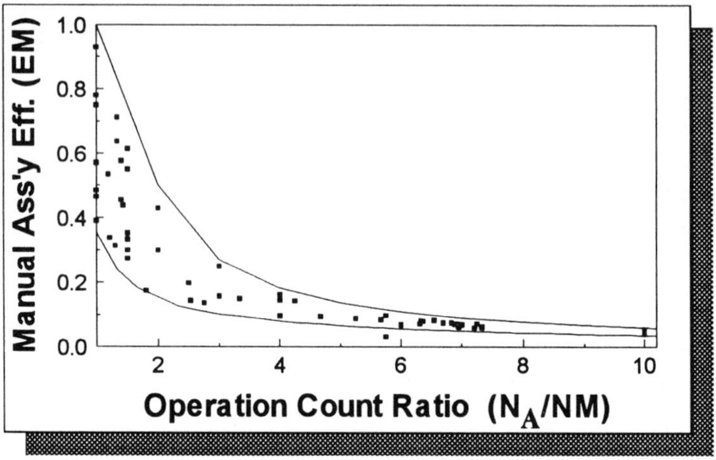

Figure 3: Bounds on DFA efficiency as a function of the Operation Count Ratio. The operation count ratio is the number of assembly operations (N_A) divided by the theoretical minimum number of parts (NM). The data points are from observations obtained by analysis performed by students for a graduate design class at Stanford. The defined bounds are empirical fits to the data based on Pareto distributions.

3.3. Assembly Defect Increases for Increased Assembly Time. Even when parts satisfy defined tolerances and requirements, defects can occur during the assembly process. One source of assembly defects is interference between mating parts. Assembly errors, such as installing a part in an incorrect position or orientation, are other sources of assembly defects. As previously mentioned, we postulated that the probability of a defect in an assembly might be related to the time required to perform an assembly operation. An analysis of several simple assemblies supported the assumption that defects would increase with assembly operation time.

As the number of features in the interface grows, the assembly time generally increases and there is a higher probability that dimension or tolerance allocation errors will occur. In addition, the longer assembly times are also related to more difficult assembly tasks. Thus, the probability of an assembly error is likely to be related to the assembly operation time for a constant level of care in the operation. Thus we postulated that the assembly time may be related to the probability that a defect would be introduced into the assembly.

Our analysis, combined with the potential of developing a method for defining a product quality strategy, prompted Motorola to review their existing data on the sources of product defects to identify the frequency of defects originating in the assembly process.[19] David Gebala, in the Advanced Manufacturing Technologies division played a key role in accumulating and organizing the data. Fifty combinations comparing defect rates to product characteristics, such as total assembly time or number of assembly operations, were examined. Of all the combinations, the average defect per operation versus the average assembly time per operation showed the strongest linear correlation (correlation coefficient "r" = 0.94).

3.4. Estimating the Product Defect Probability. In addition to the assembly errors and part interferences due to dimensional variations, parts may contain defects not associated with the interface such as material defects. These non-interface defects will be collectively referred to as part defects. The probability that a product does not contain a defect is the product of the probabilities that each operation and each part does not add or cause a defect. Given that tolerance variation, assembly errors such as misalignment, and part defects are independent variables each having small probabilities of occurrence, the combined probability of a defect free assembly (P_T) can be shown to have the following form:

$$P_T = \prod_{i=1}^{a} \{1 - c_k (t_i - t_{ideal})^k\}(1 - D_{P_i}) \qquad (2)$$

Where, P_T = Probability that the total assembly is defect free
c_k = Constant related to quality control of assembly operations
t_i = Time required to complete the ith assembly operation
k = Exponent relating defect sensitivity to the assembly time
a = Number of assembly operations
D_{P_i} = Probability that the ith part contains a defect

Equation 2 reveals five fundamental factors that can contribute to assembly defects:

1) The number of assembly operations
2) Quality Control: The capability of completing an operation without a defect (c_k), [measure of the organization's skill and effectiveness]
3) The complexity of assembly operations gauged by the time required to perform the operation
4) The number of parts (a subset of the number of assembly operations)
5) Part defect probabilities

[19] David Gebala, Correspondence, Motorola, Inc., August 7, 1992.

Each of these factors suggests specific independent strategies which must be pursued in reducing assembly defects in quality products. The best strategy to improve product quality depends upon the current level of performance in each of these respective areas. For example, if part defect rates are very low but assembly errors are frequent, further reduction in part defect rates will not substantially improve quality conformance. In this case, the maximum improvement in quality can be achieved by reducing the frequency of assembly errors.

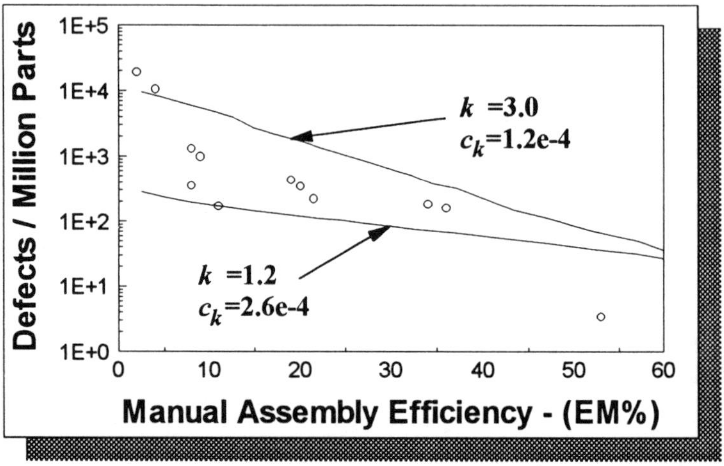

Figure 4: The average defects per million parts versus design versus assembly efficiency for various values of "k", an exponential describing the rate of increase complexity as a function of assembly time. Note the general agreement with the trend in published Motorola data.[20]

It is possible to have high assembly efficiency but poorly toleranced parts or uncontrolled processes that result in a large number of assembly defects. Conversely it is possible to have low assembly efficiency with carefully designed parts and constant assembly inspection that leads to very low defect rates such as may be encountered in a space shuttle type program. Thus, grouping the results of many different assemblers is likely to produce results that do not show a strong relationship between defects and assembly time per

---

[20] Ibid.

operation. This type of result has been observed in industry.[21] The variability in results among assemblers can be represented by a quality control constant (c_k). This constant is a measure of the ability of each manufacturer to control assembly processes to produce defect free assemblies. A value of zero for this constant would mean that all of a manufacturers assembly operations were in perfect control.

Figure 4 shows the derived relationship for the number of defects per million parts as a function of the DFA efficiency. This plot also illustrates the change in predicted defect rate as a function of the exponential value of "k." As "k" increases, the average defect probability per operation and part increases. Data published by Motorola[22] for the defect rate per part is also shown on the same chart and demonstrates that the predicted trend is appropriate.

5. Conclusions

Structured methodologies must be used as a tool which enhances global, critical thinking. The implications of design decisions should be reviewed in a broad context and critical fashion. When specific design rules are handed down from on high they tend to be accepted uncritically, precluding objective thinking by those closest to the problem.

Part count alone is not an adequate basis defining design simplicity or predicting quality conformance. The data and analysis conclusively demonstrate that an excessive emphasis on minimizing part count may result in inefficient assemblies and complex parts. While there are many benefits to be derived from minimizing part count, a blind preoccupation can lead to serious deficiencies. A superior design criterion is minimizing and simplifying assembly operations. This tends to reduce part counts and simplify part interfaces while avoiding assembly complexity which may be introduced to achieve part count reduction.

Assembly operation times follow a Pareto distribution. Using a Pareto distribution of assembly times, it has been shown that an operation count as a function of DFA efficiency can be bounded as substantiated by a large number of observations. This provides a critical link in relating product simplicity to product defect rates.

The Pareto distribution defined by the Riemann Zeta Function can be used as a powerful predictive tool in many manufacturing applications. A relationship between design efficiency and the probability of assembly defects

---

[21] James P. Womack, Daniel T. Jones, and Daniel Roos, The Machine that Changed the World, (New York: Rawson Associates, 1990), p. 93.

[22] Brannan, pp. 1-3.

has been defined which provides a theoretical basis for observations published by Motorola [9]. This relationship identifies five specific factors that must each be addressed independently to assure quality products including: a) operation counts, b) the assembly quality control of the manufacturer, c) assembly operation complexity, d) part counts, and e) part defect rates.

REFERENCES

Beckman, Sara, Draft of HP Chapter for the Manufacturing Visions Group, July 16, 1991.

Blum, Steve, Ade Mabogunje, and Brian Reilly, Design for Manufacturability Studies, 1991 Ford Explorer, INSTRUMENT PANEL ASSEMBLY, Stanford University, ME217B Class Final Report, June 20, 1991, pp. 3, 34.

Boothroyd, G., Dewhurst, P., Product Design for Assembly, Boothroyd Dewhurst, Inc., Section 2, Kingston, 1985.

Bralla, James G., Handbook of Product Design for Manufacturing, McGraw-Hill, New York, 1986, p 1-18.

Brannan, Bill, "Six Sigma Quality and DFA-DFMA Case Study/Motorola Inc.", Boothroyd & Dewhurst DFM Insight, Vol. 2 Winter 1991, pg 1-3.

CRC Handbook on Mathematics, 6th Edition, William H. Beyer editor, CRC Inc., Boca Raton, p. 78.

Daetz, Douglas, "The Effect of Product Design on Product Quality and Product Cost", Quality Progress, June, 1987, pp 63-67.

Gebala, David, Correspondence, Motorola, Inc., Advanced Manufacturing Technologies, Fort Lauderdale, FL, August 7, 1992.

Gradshteyn, I. S. and I. M. Ryzhik, Table of Integrals, Series and Products, Translated and Edited by Alan Jeffrey, Academic Press, London, 1980, pp 1072-1074.

Ishikawa, Kaoru, Guide to Quality Control, Asian Productivity Organization, White Planes, 1982, pp. 42-49.

Juran, J. M., and Frank M Gryna, Juran's Quality Control Handbook, McGraw Hill, New York, 1988.

King, Bob, Better Designs in Half the Time - Implementing QFD Quality Functional Deployment in America, Third Edition, Goal/QPC, Methuen, MA, 1989.

Pahl, G. and W. Beitz, Engineering Design - A Systematic Approach, Edited by Ken Wallace, Translated by Arnold Pomerans and Ken Wallace, The Design Council, London, 1988, pp 66-82.

Ross, Phillip J., Taguchi Techniques for Quality Engineering, McGraw-Hill, 1988.

Ross, Sheldon, A First Course in Probability-Third Edition, Macmillan, 1988, pp. 108-142.
Womack, James P., Daniel T. Jones & Daniel Roos, The Machine that Changed the World, Rawson Associates, New York, 1990, p. 93.

10

MANAGING FUNCTION CONCEPTS IN THE DESIGN PROCESS

Toshiharu Taura* & Hiroyuki Yoshikawa‡1
Department of Precision Machinery Engineering
Faculty of Engineering
THE UNIVERSITY OF TOKYO
Hongo 7-3-l, Bunkyo-ku, Tokyo 113, JAPAN

Abstract: In this paper, we discuss how and why *function concepts* are managed in the early stage of the design process. Conventional design methodologies have pointed out that *functional decomposition* is a key to the *synthetic nature* of the design process [Pahl & Beitz, 1988]. Especially in the case of original design, *function decomposition* is said to be indispensable. However, the reasons and mechanisms for *decomposing the required functions* have not been clarified. The importance of these processes has been understood only experientially. In this paper, we discuss the *function decomposing process* from a theoretical point of view. First, we outline a study on design methodology by Pahl and Beitz. Next, we present a mathematical framework for design knowledge which we developed in our previous paper. In this framework, *function* is recognized as an *abstract concept* and represented as a class in set theory. As one approach to formulating a mechanism for managing *function concepts*, we present a convergent policy for design solutions. Our model is useful for theoretical discussions on the *function decomposing* process. We show that *managing function concepts* can be viewed as one approach to an NP complete problem.

1. Introduction

Many attempts have been made to clarify the nature of the design process. Formerly, these studies were part of the general study of design methodology. Recently, this question has been approached from the viewpoint of design theory and information processing. We believe that the essential nature of design lies in the *synthetic process* of composing and decomposing.

1* Associate Professor, ‡ Professor

Previous studies of design methodology have also pointed out that *functional decomposition* is key to the *synthetic nature* of the design process [Pahl & Beitz, 1988]. While the earlier studies were descriptive, we have developed a mathematical framework for analyzing design knowledge and design processes [Taura & Yoshikawa, 1982]. The mathematical models are based on the General Design Theory developed by Yoshikawa [1982], and are useful for theoretical discussions of the *function decomposing process*. In our previous paper, we introduced a *metric* for design knowledge and formulated a *function decomposing process* [Taura & Yoshikawa, 1992]. In this paper, we discuss the reasons and mechanisms for the *function decomposing process* not only from an experiential but also from a theoretical point of view.

2. Design Process in Design Methodology

In this section, a study on design methodology by Pahl and Beitz is outlined. The following is a quotation from their study [Pahl & Beitz, 1988].

"The design process is composed of four steps: clarification of the task, conceptual design, embodiment design and detail design. Figure 1 shows this process step by step. The *conceptual design phase* involves the establishment of *function structures*, the search for suitable *solution principles* and their combination into *concept variants*. The *conceptual design phase* consists of several steps none of which may be skipped if the most promising solution concept is to be reached. In the subsequent embodiment and detail design phases it is extremely difficult or impossible to correct fundamental shortcomings of the concept. A successful solution is more likely to spring from the choice of the most appropriate principles than from exaggerated concentration on the finer points."

Functions are usually defined statements consisting of a verb and a noun, for example 'increase pressure', 'transfer torque' or 'reduce speed'. They are derived from the conversions of energy, material and signals. So far as is possible, all these data should be accompanied with specifications of the physical quantities.

An *overall function* can often be divided directly into identifiable *sub-functions* corresponding to sub-tasks. The relationship between *sub-functions* and overall function is very often governed by certain constraints, inasmuch as some *sub-functions* have to be satisfied before others. On the other hand it is usually possible to link sub-functions in various ways and hence to create variants. In all such cases, the links must be compatible. The meaningful and compatible combination of *sub-functions* into an *overall function* produces a so-called *function structure*, which may be varied to satisfy the *overall function*.

The object of breaking down a complex *function* is to determine *sub-functions* that will facilitate the subsequent search for solutions, and combine these *sub-functions* into a simple and unambiguous *function structure*.

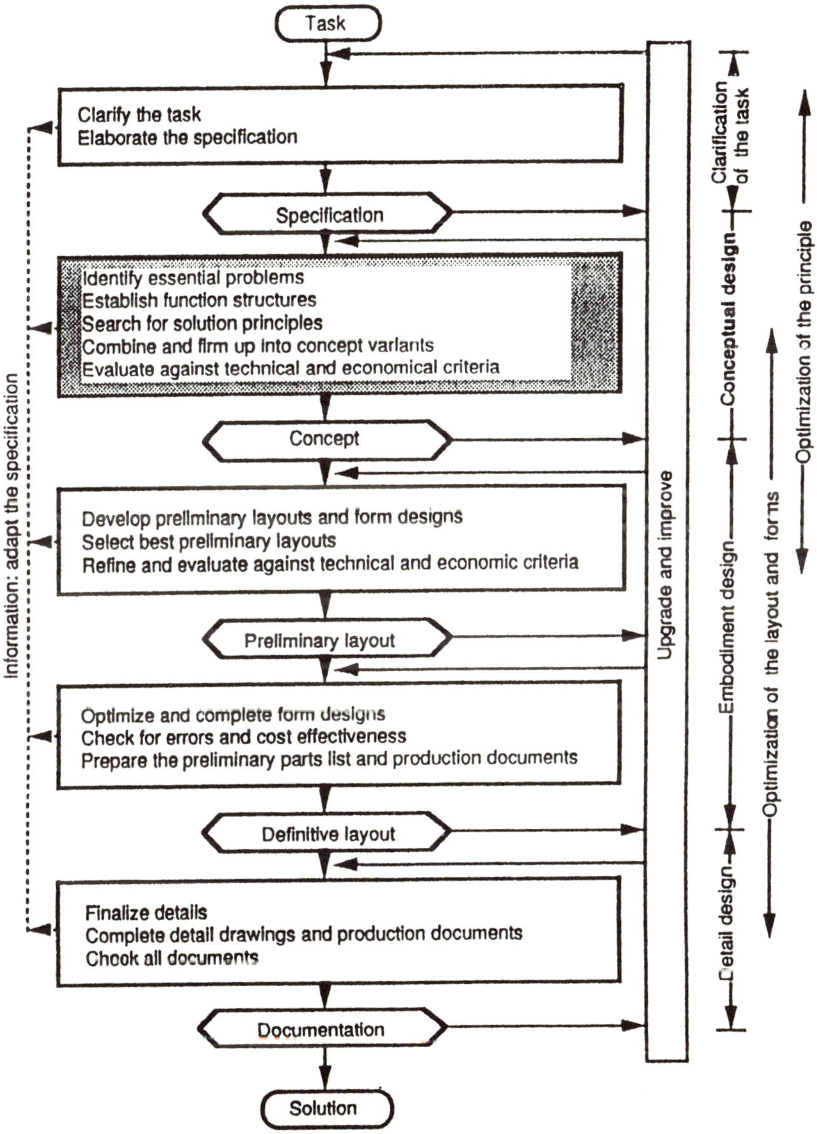

Figure 1: Steps of the design process by Pahl and Beitz

The optimum method of breaking down an *overall function*--that is, the optimum number of *sub-function* levels and also the number of *sub-functions* per level--is determined by the relative novelty of the problem and also by the method used to search for a solution.In the case of original designs, neither the individual *sub-functions* nor their relationships are generally known. In that case, searching for and establishing an *optimum function structure* constitute some of the most important steps of the conceptual design phase.In the case of adaptive designs, on the other hand, the general structure with its assemblies and components is much better known, so that a *function structure* can be obtained by the analysis of the product to be developed. Depending on the special demands of the specification, that *function structure* can be modified by the variation, addition or omission of individual *sub-functions* or by changes in their combination. A further advantage of setting up a *function structure* is that it allows a *clear definition* of existing sub-systems or of those to be newly developed, so that each sub-system can be dealt with *separately*. If existing assemblies can be assigned directly as complex *sub-functions*, the subdivision of the *function structure* can be discontinued at a fairly high level of complexity. In the case of new assemblies or those requiring further development, however, the division into *sub-functions* of decreasing complexity must be continued until the search for a solution seems promising. Establishing a function structure facilitates the discovery of solutions because it simplifies the general search for them and also because solutions to *sub-functions* can be elaborated separately. As a result, *function structures* may save a great deal of time and money.

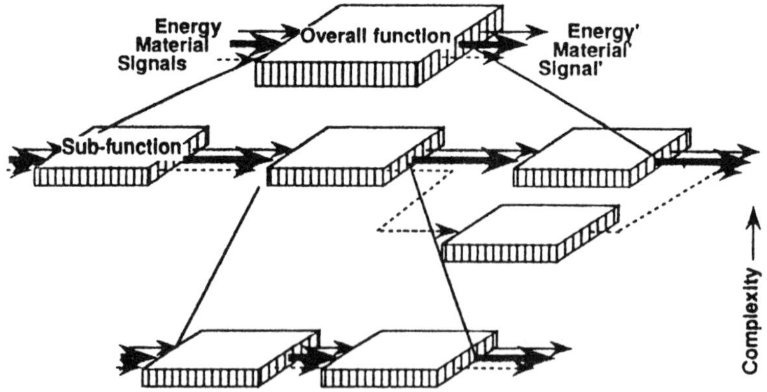

Figure 2 : Function structure by Pahl and Beitz

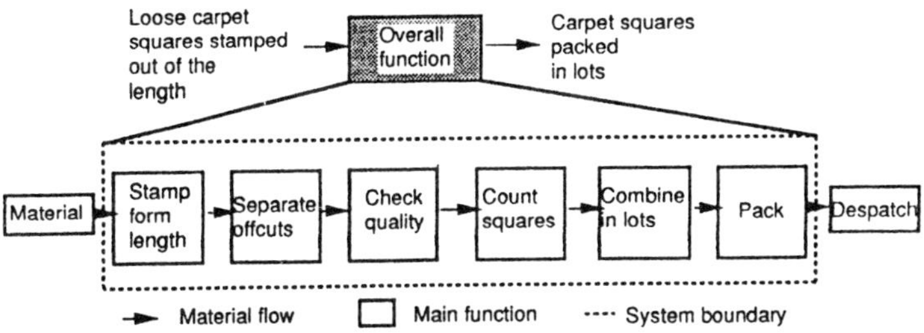

Figure 3: An example of Function structure by Pahl and Beit

Individual *sub-functions*, originally represented by black boxes, must now be replaced with more concrete statements. *Sub-functions* are usually fulfilled by physical processes-- nearly all engineering solutions are based on physical phenomena. Physical processes are based on *physical effects*. A *physical effect* can be described quantitatively by means of the physical laws governing the physical quantities involved. Several physical effects may have to be combined in order to fulfill a *sub-function*. For example the operation of a bimetal strip is the result of a combination of two effects, namely thermal expansion and elasticity. If, in concrete cases, these effects are assigned to a *sub-function*, we obtain the *physical principle* of that *sub-function*.

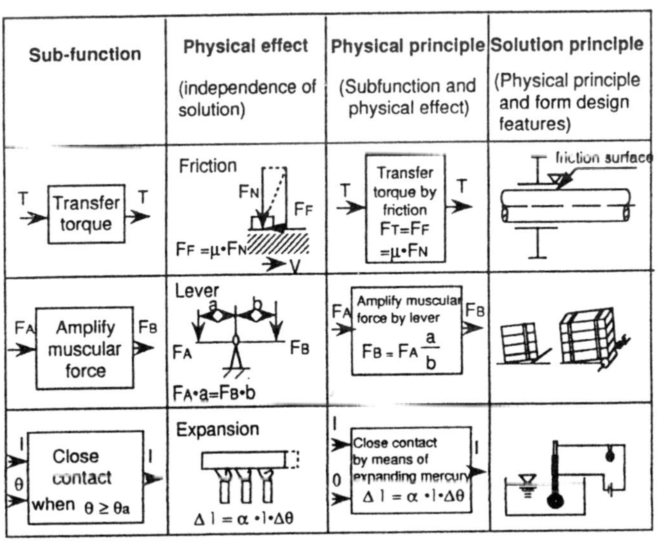

Figure 4: Fulfilling sub-functions by solution principles by Pahl and Beitz

Solution principles have to be found for the various *sub-functions* and these principles must eventually be combined. A solution principle must reflect the *physical effect* needed to fulfill a given function. The following methods are useful in the search for solutions.

- Convention aids: literature search, analysis of natural systems, analysis of existing technical systems and analogies;
- Methods with an intuitive bias: brainstorming, synectics etc.;
- Methods with a discursive bias: systematic study of physical processes, systematic search with the help of classification schemes and the use of design catalogues;

The main problem with such combinations is ensuring the physical and geometrical compatibility of the *solution principles* to be combined, which in turn ensures the smooth flow of energy, material and/or signals. A further problem is the selection of technically and economically favorable combinations.

3. General Design Theory

In General Design Theory, [Yoshikawa, 1982], the design process is modeled as a mapping from the *function space* to the *attribute space.* According to General Design Theory, an entity is recognized as an *entity concept* and represented as an element in *set theory.* The *attributes* of an *entity* are its properties (physical, chemical, mechanical, geometrical etc.) which can be scientifically observed. In this paper, *behavior* is defined as any *change* in *attribute. Attribute* and *behavior* are facts that can be known. *Function concepts* are concepts which are extracted from the *attributes* or *behavior.* A *function concept* is represented as a class, which includes the entities that manifest this function, in *set theory.*

Function concepts, which are extracted by paying attention to the assembly's *attributes* or *behavior* as a unit, are called *total function concepts* and the space, the elements of which are assemblies and the classes of which are *total function concepts*, is called a *total function space.* In contrast, *function concepts* which are extracted by paying attention to the components are called *partial function concepts* and the space, the elements of which are the compositions of components and the classes of which are *partial function concepts*, is called a *partial function space.* A mapping from a *total function space* to a *partial function* space is called *function decomposition.*

An attribute concept or *behavior concept* is also a class, and includes the entities that have this attribute or behavior. *A component concept* is also a class in *set theory*, and includes the entities that can be recognized as the same component from the viewpoint of its properties or shapes.

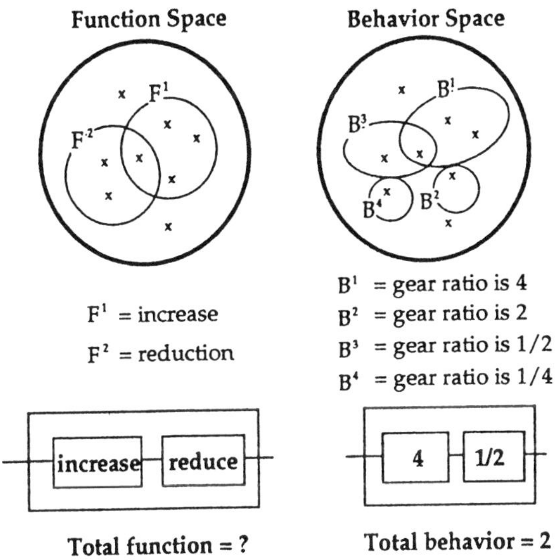

Figure 5: Difference between Function and Behavior

In General Design Theory, *abstract concept* is also a class. *Function concept, attribute concept, behavior concept* and *component concept* are recognized as types of *abstract concepts*. The differences among them depend on the degree of abstraction. To illustrate the differences, from a semantical point of view, consider the example in Figure 5. In this example, it is possible to infer the *behavior* of an assembled object from that of its components. However, it is impossible to infer the *function concept* of an assembled object from that of its components. It can be extracted from an assembled object's *behavior*. Thus, it is assumed that the relation between *total function concept* and *partial function concept* becomes weaker as the *function concept* becomes more abstract. In other words, *function concept* can be defined at such a level of abstraction that it becomes impossible to infer the *function concept* of an assembled object from that of its components.

The above discussion suggests the following issues. First, how can such a highly abstracted concept be operated? Next, why is such a highly abstracted concept used in the design process if the above model does not fit in with an actual design?

In this section, a *metric* is introduced into design knowledge as one approach to operating a highly abstracted concept. Also, the rationale to use a highly abstracted concept is discussed from the viewpoint of cognitive efficiency and complexity of computation.

3.1. Total Function Space and Partial Function Space. In this sub-section, the spaces on which the *functional operation* can be mathematically modeled are defined. The discussion begins with defining a fundamental space.

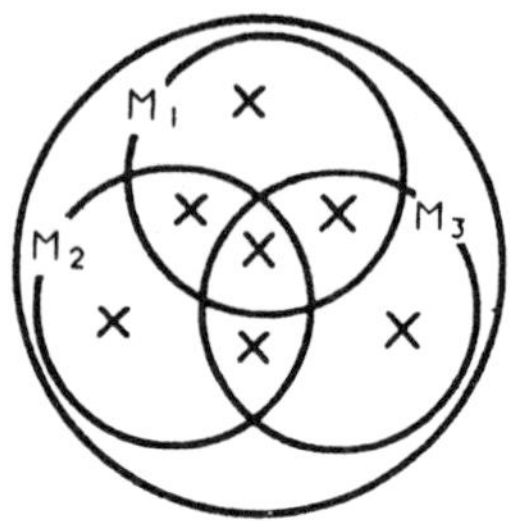

Figure 6: Ideal function concept set

Def.1: An *ideal function concept set (M)* is defined as a family of sets introduced into a subset (S_0) of the *entity concept* set and has the following properties.

(1) The number of *function concepts* (subset of S_0) which constitutes the *ideal function concept set* as its classes is at most countable. **M** = {M_i _ M_i _ S_0, i _ _, _: at most countable}

(2) As long as there is a finite number of *function concepts*, an intersection of these concepts will never be null. (M_{n1} _ M_{n2} _ ...... _ M_{ns}) _ _

(3) For any *function concept*, there exists an element of S_0 which belongs to this *function concept* and does not belong to any other *function concept.*

(4) For every two elements of S_0, there exists M_i _ **M** which includes one element without including the other.

The properties of (2)-(4) are explained intuitively as a functional structure which can discriminate the maximum number of elements. Among these four properties, property (3) is very important for developing a *partial function space.*

We now describe a framework for *functional* operation.

Def.2: A *total function concept* is a *function concept* which an *entity concept* manifests as a whole. An *ideal total function concept set* ($\mathbf{M_t}$) is an *ideal function concept set* (**M**) which is obtained by paying attention to the *total function concept.*

Def.3: A *total function space* (S_1, $\mathbf{M_t}$) is defined as a space, the elements of which are a subset of the *entity concept set* and the *ideal total function concept set* ($\mathbf{M_t}$).

Next, we consider the *function concept* in terms of the *parts.*

Def.4: For any function concept M_t^q which belongs to an *ideal total function concept set* ($\mathbf{M_t}$) there is an element which belongs to the *function concept* (M_t^q) and does not belong to any other *function concept* of $\mathbf{M_t}$. This *entity concept* is called the *representative design component* (e_q). The set of *representative design components* is denoted by E.

Def.5: A *design entity concept set* (S_d) is defined as the *power set* of a *representative design component set.* An element of S_d is called a *design entity concept.*

A *representative design component* is similar to a component listed in a design handbook and a *design entity concept* is similar to a composition made up of these components.

Theorem.1: The *design entity concept set* (S_d) and the base set (S_1) of the *total function space* have a one-to-one correspondence.

Proof: First, note that the potency of the set (class) which constitutes the *ideal total function concept set* (M_t) is finite (_$\mathbf{M_t}$_ = **m**). From the properties of (2) and (3) in Def.1, set S_1 can be divided into 2^m - 1 regions, by the *m function concepts.* The properties of (2) and (4) in Def.1 indicate that there is only one *entity concept* in every region. The potency of S_1 is therefore 2^m - 1. On the other hand, the potency of the *design entity concept set* (S_d) is equal to the potency of the *power set* of *m* number of elements other than null; i.e., 2^m - 1. These two sets have a one-to-one correspondence because both of the potencies are the same. Next, we consider the case when _$\mathbf{M_t}$_ = _ (_ means that the potency of the set is countable). Employing arguments used for the finite potency case, we can show that the potency of both S_d and S_1 is 2_. Thus set S_d and set S_1 have a one-to-one correspondence.

This theorem is intuitively explained by the fact that the *entity concept set* can be formed as a composition of the *representative design component set.*

Def.6: For every *representative design component* (e_q), a *partial function concept* M_p^q is defined as a subset of the *design entity concepts* which include this *representative design component* (e_q). The *partial function concept set* is called an *ideal partial function concept set* ($\mathbf{M_p}$).

Strictly speaking, the *partial function concept* is not a class obtained by paying attention to each *representative design component*, but one obtained by paying attention to the (total) *function concept* manifested by each *representative design component.* Thus Def.1 and Def.6 both generate the same classes. Def.6 is defined in order to simplify the mathematical expression.

Def.7: The *partial function space* (S_d, $\mathbf{M_p}$) is defined as the space, the elements of which are the *design entity concept set* (S_d) and the classes of which are the *ideal partial function concept set* ($\mathbf{M_p}$).

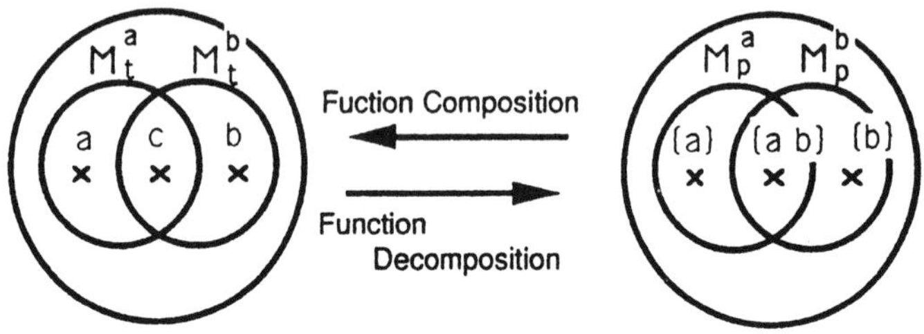

Figure 7: Total function space and partial function space

3.2. A Mathematical Model of Functional Operation and Conceptual Design Process

Def.8: *Function composition* is the mapping of a *design entity concept* in a *partial function space* onto an *entity concept* in a *total function space*. *Function decomposition* is the mapping of an *entity concept* in a *total function space* onto a *design entity concept* in a *partial function space*.

The discussion on the conceptual design process model starts from the following definition, which was presented Tomiyama & Yoshikawa [1985].

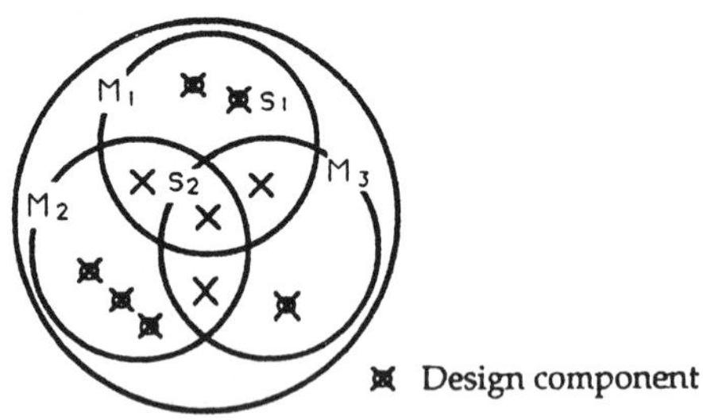

Figure 8: Design components

Def.9: *Designing* is designating a domain in the *attribute space* corresponding to the domain specified by the specifications of the *abstract concept space*. In this paper, attention is focused on a space where the design solutions are represented by the component. In other words, *attribute (behavior)* is assumed to be dealt with by means of the components.

Def.10: In every *function concept* of an *ideal function concept set* (**M**), the elements of the *entity concept set* which belong to this *function concept* and which do not belong to the other *function concepts* are called *design components.*

Def.11: A *component space* is defined as a *discrete topological space* the elements of which are a *power set* of a *design component set.*

Def.12: *The conceptual design process* is defined as the composition of the following operations.

phase 1: to designate a domain on the *partial function space* corresponding to the domain on the *total function space.*

phase 2: to designate a domain on the *component space* corresponding to the domain on the *partial function space.*

phase 3: to designate a domain on the *attribute space* corresponding to the domain on the *component space.*

phase 4: to designate a domain on the *total function space* corresponding to the domain on the *attribute space.*

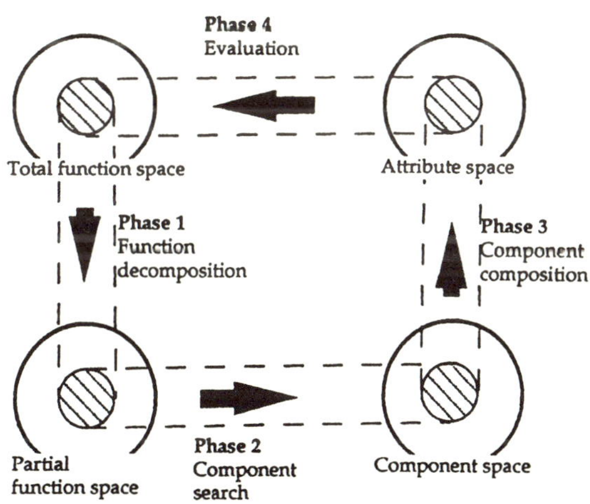

Figure 9: A mathematical model of the conceptual design process

Each *phase* in the above definition refers to the following operation.

Phase 1 is to *decompose* a given specification (*function concept*).

Phase 2 is to search for the *design components.*

Phase 3 is to compose the *design components.*

Phase 4 is to *evaluate* the candidates for the design solution (***entity concept***).

4. Nature of Function Decomposition

4.1. Design Process in Design Methodology and in General Design Theory. Let us analyze the design process model by Pahl and Beitz using the above theoretical framework. In Fig. 10, the *conceptual design process* developed by Pahl and Beitz is compared to the framework in General Design Theory. Here, some terms are seen to correspond to each other as follows.

Overall function in the methodology of Pahl and Beitz corresponds to *total function concept* in General Design Theory. The *functions* are defined statements consisting of a verb and a noun, for example, "increase pressure", "transfer torque" or "reduce speed". In the example of the theory which is shown later, however, *the total function concepts* are represented as "move", "increase speed" or "reduce speed". Both forms of expression are found to be similar.

Sub-function in the methodology corresponds to *partial function concept* in the theory. While the methodology implicitly defines *sub-functions* as functions of their components and distinguished from sub-task, the theory explicitly defines *partial function concepts* as those of their components. Either way, the components play an important role in *decomposing functions.*

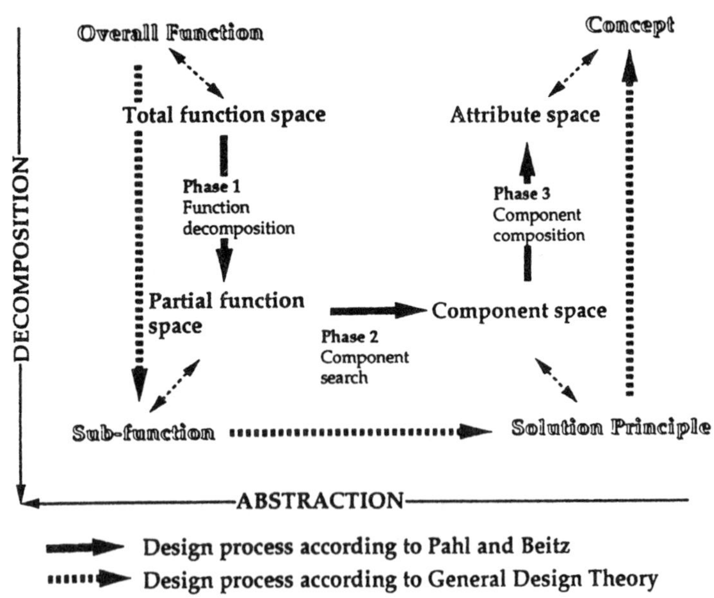

Figure 10: Design Process in Methodology and in Theory

Solution principles in the methodology correspond to the *component's behavior* in the theory. In the methodology, *solution principles* are built up of *physical principles* and *form design features*. The theory defines *behavior* as any change in *attribute*, the properties (physical, chemical, mechanical, etc.) of which can be scientifically observed. And *attribute (behavior)* is assumed to be dealt with by using the components.

Furthermore, as shown in section 3, it is assumed that the *behavior* of an assembled object can be inferred from that of its *components*. Therefore, dealing with the *components* as a unit in inferring *behavior* in the theory seems to correspond to the *solution principle* in the methodology.

The *design solution* of the *conceptual design process* in the methodology corresponds to the assembly of *components* in the theory. In the methodology, the *design solution* of the *conceptual design process* is the combination of *solution principles*. These are thought to correspond to the *assembly of components* in the theory.

As is mentioned above, the theory captures the characteristics of each phase described in the design methodology by Pahl and Beitz.

4.2. Nature of Conceptual Design Process. Using the framework in Fig. 10, we can discuss the nature of the conceptual design process theoretically.

First, let us discuss the relation between the upper side and the lower side. This relation involves the *function decomposing process* and the *component composing process*. Then, one axis of the conceptual design process represents the *composition* and *decomposition processes*.

Next, let us discuss the relation between the left side and the right side. As mentioned in section 3, the difference between the function space and the attribute (behavior) space can be discussed using the strength of topology, something like the size of the subset in set theory. According to the previous study, the topology of the attribute space is stronger than that of the function space under the condition that designing is possible [Yoshikawa, 1982]. The above discussion also suggests that *component space* can be regarded as a kind of *attribute (behavior) space* and that the semantical relation between the left side and the right side indicates the degree of abstraction. Then, the other axis of the conceptual design process represents *abstraction*.

As a result, the *conceptual design process* is considered to be built up of the processes of *(de)composing* and *abstraction*. Observe that the task of designing is simply to find design solutions which manifest the required specifications (function). Thus, for the purpose of designing itself, neither the *function decomposition process* nor the *component search process* is needed. Furthermore, it is difficult to deal with such highly abstracted concepts as the *function concepts*. Then, it is natural to believe that there should be some purposes for such abstract concepts and some mechanisms to operate them in a designer's mind. In the next section, the mechanism to operate the *function concepts* is discussed.

5. A Convergent Policy for Design Solution

The basic idea of searching for design solutions is the *paradigm model* proposed in a previous paper [Yoshikawa, 1982]. The *paradigm model* is as follows.

Given a specification $T = T_1 \_ T_2 \_ \cdots\cdots T_n$. Assume that a solution, s_1, is proposed, which satisfies

$$s_1 \_ T(s_1) = T_1 \_ T_2 \_ \cdots\cdots \_ T_k \_ k{+}1 \_ k{+}2 \_ \cdots\cdots \_ \mathrm{n}.$$

In this case, s_1 satisfies only a part of the specification. Next, assume the next proposal, s.

$$s_2 \_ T(s_2) = T_1 \_ T_2 \_ \cdots\cdots \_ T_k \_ T_{k+1} \_ k{+}2 \_ \cdots\cdots \_ \mathrm{n}.$$

This also only satisfies a part; however, it does represent an improvement over s_1. In this case, the move from s_1 to s_2 will satisfy the characteristics of a directed series of points. Let us call these points, which are the proposed solutions, *paradigms*.

An example of the *paradigm model* is shown in Fig. 11. The nature of the design knowledge and the design process in the *paradigm model* is discussed using Fig. 12. First, the design knowledge, which is used while designing in the *paradigm model*, is found to be a *partial function space*. For example, the right side picture in Fig. 12 is a *partial function space*, which seems to be organized from the precedents. Next, the design process in the *paradigm model* can be discussed in this *partial function space*. For example, the design process of the above example is illustrated by the arrow in the right side picture in Fig. 12. Mathematically, such a movement of *paradigms* shown in Fig. 12 is the neighborhood searching in a *topological space*.

The problem is the methodology applied to move concretely from s_1 to s_2. In the example shown in Fig. 11 and Fig. 12, a *partial function concept* of a component of a design object is also recognized as a *total function concept*. Then, the methodology applied to move from s_1 to s_2 can be easily determined. But this is not usual. A method for refining and reconstructing a design object in order to move the object when searching for a design solution has not generally been required.

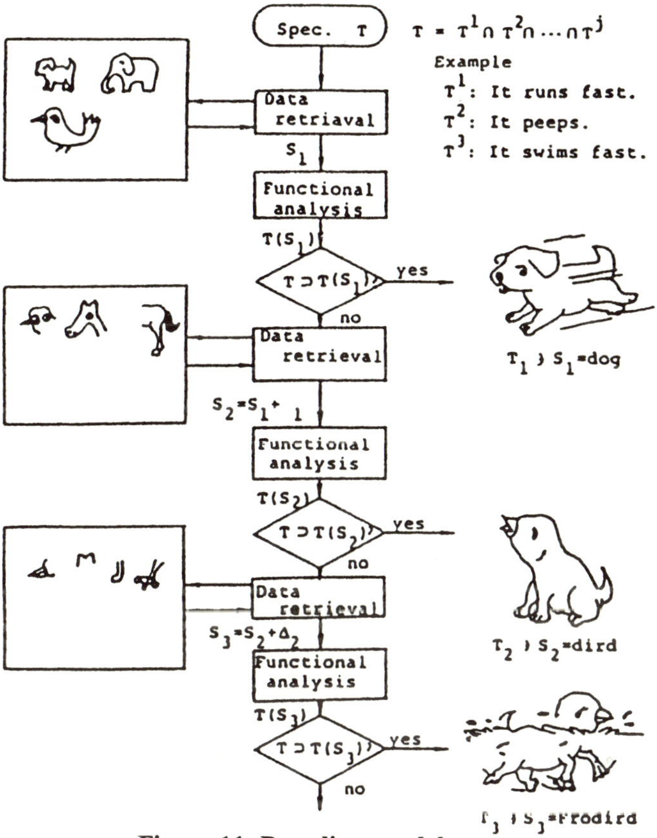

Figure 11: Paradigm model

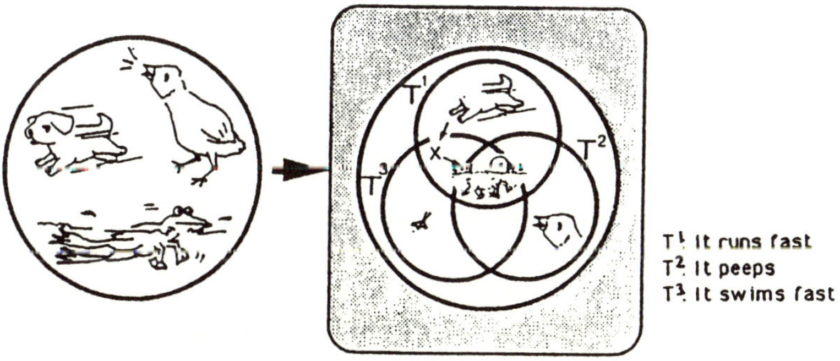

Figure 12: Design Knowledge and Design Process in this study

5.1. A Metric Space for Design Knowledge. A key idea to offer a driving force for the paradigm model is to introduce a metric into design knowledge. In this sub-section, the theory in section 3 is extended in order to find a mathematical nature for driving the design process. First, a mathematical process to improve the *ideal function concept set* into a *metrized topological space* is determined.

Def.13: To *structure* an *ideal function concept set* is to generate a *topological space* (S_0,), the classes of which consist of _, S_0, every form of _ (M_{n1} _M_{n2} _ ...... _ M_{ns}). Here, _ refers to an intersection of a finite number of classes, while _ refers to a union of a finite or infinite number of classes.

Def.14: To *organize a structured ideal function concept set* (S_0,) is to generate a *compact Hausdorff space* by adding some classes (') to the space (S_0,).

Def.15: A *metrized total function space* (S_1,$\mathbf{T_t}$) is defined as a space obtained by structuring and organizing *a total function space* (S_1, $\mathbf{M_t}$).

Def.16: The *metrized partial function space* (S_d,$\mathbf{T_p}$) is defined as a space obtained by *structuring* and *organizing a partial function space* (S_d, $\mathbf{M_p}$).

From the process of generating a *compact Hausdorff topological space* as described in the above definitions, the following process can be imagined. He/she begins to discriminate some entity concepts by applying some *function concepts*. As he/she grows up, more *function concepts* are learned. Then all knowledge is classified again. After these processes have been repeated several times, the ideal described in Def.1 is obtained. Next, the ability to discriminate expands and becomes *structured* knowledge, the *topological space* defined in Def.2. Finally, the nature described in Def.15 is obtained and a *metric* is introduced into the *structured* knowledge.

The following theorem about the total function space is derived.

Lemma 1: A *compact Hausdorff space* is the *normal space* (T_4).

Lemma 2: (Urysohn) A *topological space* X is *normal*. Then, for every two mutually disjointed, closed subsets A, B on X, there exists a *real continuous function* (f: X _ R) which satisfies the following conditions.

(1) $f(x) = 0$ for all x _ A and $f(x) = 1$ for all x _ B

(2) for any x _ X, 0 _ f(x) _ 1

Theorem 1: In a *metrized total function space* there is a *metric* between two different *entity concepts*.

Proof: Lemma 1 and Def.3 indicate that a *metrized total function space* is a *normal space*. Meanwhile, from Lemma 2, a *real continuous function* (f) is obtained on which the two *entity concepts* can be mapped.

In this study, to be able to introduce the *metric* into a space means to be able to obtain a *real continuous function* in Lemma 2.

Let us consider the nature of *function composition* and *function decomposition*. *Function composition* and *function decomposition* can be defined in the spaces which have been neither *structured* nor *organized*. However, an inkling into the nature of *function composition* and *function decomposition* can be obtained by introducing a *metric* into the knowledge space.

Lemma 3: The following is a necessary and sufficient condition so that a function (f) on X is a *real continuous function*.
Let x_0 be any point on X. Then, for any real positive number _

$$f^{-1}((f(x_0) - \_, f(x_0) + \_)) = \{ x \_ x \_ X, \_ f(x) - f(x_0) \_ < \_ \}$$

is in the neighborhood of x_0 on X.

Theorem 2: Let us focus on an *entity concept* $s_1{}^1$ and two closed sets, neither of which includes this *entity concept* on a *metrized total function space*. Let us also focus on a *design entity concept* $S_d{}^1$ and two closed sets, neither of which includes this *design entity concept* on a *metrized partial function space*. Furthermore, let *nearness* be evaluated in a *real line* defined in Lemma 2. Then, for the *entity concept* $s_1{}^1$ on the *metrized total function space*, a *metrized partial function concept* with a *nearness* to $s_1{}^1$ which is arbitrary can be obtained. Also, for the *design entity concept* $S_d{}^1$ on the *metrized partial function space*, a *metrized total function concept* with a *nearness* to $S_d{}^1$ which is arbitrary can be obtained.

Proof: - See Yoshikawa (1982).

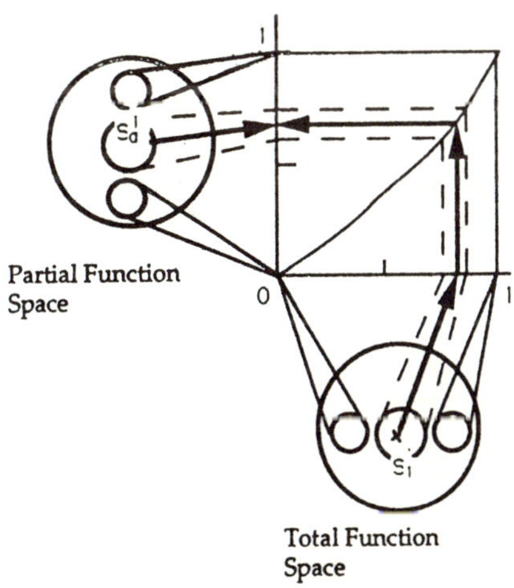

Figure 13: Relation between the total function space and the partial function space

Theorem 2 relates *entity concepts* which are *near* each other on the *metrized total function space* to *design entity concepts* which are *near* each other on the *metrized partial function space*. We assume here that the design knowledge in an actual design process is of the same nature as *Theorem 2*.

Hypothesis 1: In *function decomposition*, elements which are *near* each other on the *total function space* are mapped onto elements which are *near* each other on the *partial function space*. The inverse is also valid.

5.2. Functional Similarity between Objects. The key idea in *Hypothesis 1* which offers a driving force for the *paradigm model* is that two design objects which are *similar* in their *partial function structure* may manifest *similar total function concepts*. This hypothesis is explained using the example in Fig. 14. Let us consider the relation between a *total function concept* and a *partial function concept* in the example in Fig. 14. First, let us examine the *total function concept* of these machines in terms of "motion". Both of these machines are found to manifest the same *total function concepts* such as a stroke-increase *function* and a rectilinear motion *function*. Next, let us examine the partial function concept of these two machines. First, MACHINE(1) is understood to be composed of some components, *cyl, lever* and *bar*. Suppose that the *function concept set* {*move, rotate*} is recognized. In that *case, cyl* of MACHINE(1) will belong to *move*. Mathematically, *cyl* is an element of the *move* class. The other components, *lever* and *bar*, are also found to belong to *rotate* and *move*, respectively. Furthermore, MACHINE(1) is said to be composed of the *partial function concepts* {*move, rotate, move*}. As in the case of MACHINE(1), MACHINE(2) is composed of *cyl, pinion* and *rack*, and it is also composed of the *partial function concepts* {*move, rotate, move*}. Here, one will notice that both *partial function concepts* are the same, though they are composed of quite different mechanical components. As a result, between two machines which manifest the same *total function concepts* but are composed of quite different mechanical components, the *similarity* in their partial structure is obtained by paying attention to their *partial function concepts* .

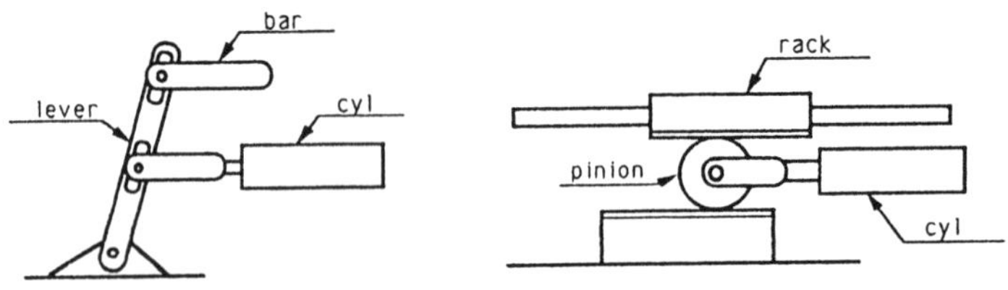

MACHINE(1) **MACHINE(2)**

Figure 14: An example of functional similarity between machines

6. An Algorithm to Generate Design Solutions

Extending the theory developed in the previous section, an algorithm to find the design solutions for the preceding examples has been proposed in our previous study [Taura & Yoshikawa 1992]. This algorithm is a generate-test method, distinct in that candidates for design solutions are refined in the *partial function space* by paying attention to the *metric* as the first meaning.

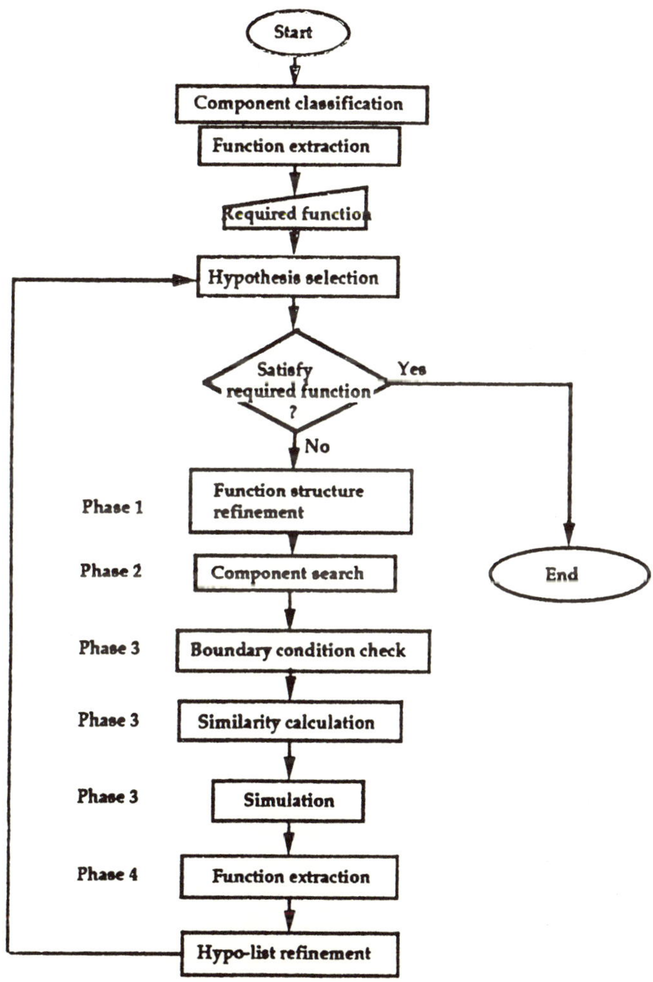

Figure 15: Algorithm for the generation of design solutions.

6.1. Basic Plan. The nature of the *convergence policy* proposed in the previous study is as follows.
(1) Start from a previous example.
(2) Evaluate candidates for the design solution in the *total function space.*
(3) Refine candidates in the *partial function space* by paying attention to the *metric.*

6.2. An Algorithm to Generate Design Solutions. The algorithm to generate the design solutions is shown in Fig. 15. The number of each *phase* in this figure corresponds to the number of each *phase* in section 3.2. Some modules are explained.

- *Component Classification Module*

With this methodology, before the design solutions are generated, all the components are classified by applying the rules described in the *knowledge base*, and the *component concepts* are generated.

- *Function Extraction Module*

As mentioned in section 3, the *total function concept* of a newly assembled object obtained by changing its components can be extracted from the assembly's *attributes* or *behavior* by applying given rules.

On the other hand, *function extraction* can be carried out from a variety of *views.* Each *view* corresponds to a *function space* which comprises the *function concepts* that are related to each other. That is, a *total function space* or a *partial function space* is regarded as a *function space* recognized from one *view*, while the actual design knowledge is believed to be a composition of *function spaces.* In addition to this, the *relationships* among the *partial function concepts* are regarded as a meaningful part of the actual design knowledge. Considering *view* and *relationship*, a *partial function space* can be regarded as a *structure.* This *structure* is called a *partial function structure.*

- *Function Structure Refinement Module*

The selected *hypothesis* can be refined by changing its *partial function structure.* In this module, the *hypothesis* is operated on a *partial function space.* In this paper, the order of *partial function concepts* defines the *relationship* among these *partial function concepts.* The *function structure* can be refined as follows.

(1) The *partial function structure* can be refined in every *view.*
(2) In refining *partial function concepts*, the *neighborhood* of the unrefined *hypothesis* in the partial function space can be searched. The available concrete methods are as follows.
   - Search the *neighborhood* by *adding* one *partial function concept.*

- Search the *neighborhood* by *changing* one *partial function concept.*
- Search the *neighborhood* by *deleting* one *partial function concept.*
- Search the *neighborhood* by *changing* the order of the *partial function concepts.*

(3) The refined *partial function concepts* obtained in every *view* are composed according to each operation, *adding, changing* and *deleting.* As the result of this composition, refined *partial function structures* can be obtained.

6.3. Example. In the example, *conceptual design* to search for mechanical component structures which satisfy the given specifications has been attempted. As for the *relationship* among the *partial function* concepts of the machine, the energy flow becomes the focus. The series connection from input to output is chosen as the subject.

6.3.1. Data Base. In the *data base,* the *attributes* and *behavior* of the machines in Fig. 16 and their components are described.

In this example, the *total function concepts* of the assembled machine and the *partial function concepts* of its components are extracted from three *views.* The first *view* is whether the output motion is *move* or *rotate.* The second *view* is whether or not the speed is *reduced* at the end of the motion. The third *view* is whether or not the speed of output is *increased* against that of the input.

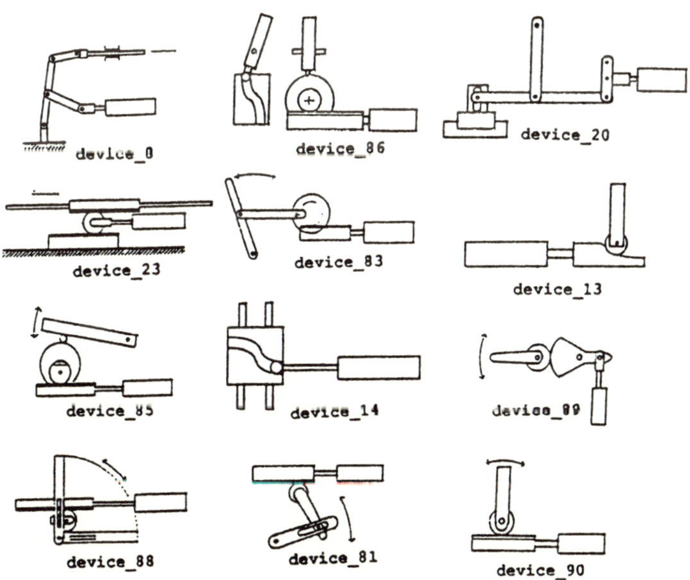

Figure 16: Machines described in the data base

```
(DEVICE-88 (INPUT     (TYPE AIR)))
           (OUTPUT (TYPE (MOTION))
                   (TRANS (PIN))
                   (DIR ((1 0 0)(1 -1 0)(0 -1 0)))
                   (SP-PAT (1 4 1)))
           (FUNC    (BY-BEHAVIOR (ROTATE)(SP-RED-ENDS)(SP-INCREASE)))
           (COMPOSED-OF
                   (PART (DEVICE-88-1 DEVICE-88-2 DEVICE-88-3
                          DEVICE-88-4 DEVICE-88-5))
                   (FUNC ((MOVE)(NIL)(NIL))
                         ((MOVE)(NIL)(NIL))
                         ((ROTATE)(NIL)(NIL))
                         ((ROTATE)(NIL)(SP-INCREASE))
                         ((ROTATE)(SP-RED-ENDS)(NIL)))))
```

```
(DEVICE-88-2
         (INPUT   (TYPE (MOTION))
                  (TRANS (PIN))
                  (DIR ((1 0 0)(1 0 0)(1 0 0)))
                  (SP-PAT (1 1 1)))
         (OUTPUT  (TYPE (MOTION))
                  (TRANS (RACK))
                  (DIR ((1 0 0)(1 0 0)(1 0 0)))
                  (SP-PAT (1 1 1)))
         (NAME    (ORIGINAL (RACK))
                  (BY-BEHAVIOR (RACK16)))
         (FUNC    (BY-BEHAVIOR (MOVE)(NIL)(NIL))))
```

[shaded] is inserted by the system using the knowledge base

Figure 17: An example of description in the data base

In the slot *INPUT*, information about the input of a machine is described prior to starting the system. In the slot *OUTPUT*, information about the output is described. In these slots, the *behavior* at three different times, the *beginning*, *middle* and *end* of the motion, is described. For example, the directions of the motion in a 3-dimensional space are described in slot *DIR* and the speed at each moment is described in slot *SP_PAT*. The speed in the *middle* of the *input* is defined as "1" as the standard.The *function concepts* are extracted using the rules provided in the *knowledge base*. The extracted *function concepts* are inserted into slot *BY-BEHAVIOR* in slot *FUNC*.

6.3.2. Knowledge Base. The dynamic knowledge peculiar to the design objects has been written into the *knowledge base*. In this system, (1)information on *extracting* the *function concepts*, (2)information on *classifying* the components, (3)information about *boundary conditions*, and (4)information about *simulations* are provided.

Regarding the information on *extracting* the *function concepts*, for example, there is information indicating that if all the directions at the *beginning*, *middle, end* in slot *DM* are the same, then *move* will be written in.

Regarding the information about *boundary conditions*, boundary conditions about the ends of a component and the connection between two components which are next to each other are written in. For example, information which can be used to determine whether the directions of the *OUTPUT* in the previous component can be superimposed on the directions in the *INPUT* in the next component by rotating the axes is written in.

Regarding the information about *simulations*, the directions and the speeds of an assembled machine are simulated. For example, the following information has been written to *simulate* the direction of motion. First, the rotating angle of the axes is calculated to superimpose the directions of the *OUTPUT* from the previous component on the directions of the *INPUT* in the next component. Next, the *OUTPUT* directions in the next component are calculated by rotating the axes using the obtained angle.

6.3.3. Example. Using the *data base* and the *knowledge base*, a search was attempted for mechanical component structures which manifest the *move function* as the first *view*, the *speed reduction function at both ends* as the second *view*, and the *speed increase function* as the third *view*. The action of the system is as follows.

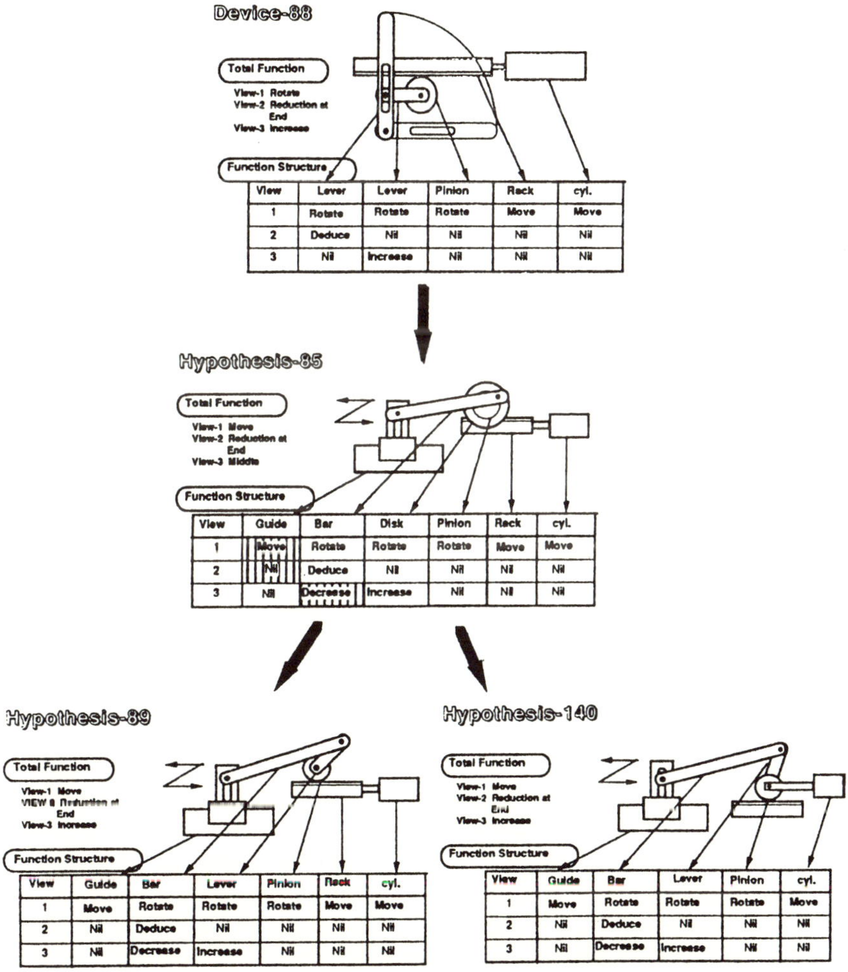

Figure 18: Transformation to reach the design solution

First, the system starts from *DEVICE_20*. *DEVICE_20* is one of the precedents most similar to the required specification. However, the system fails to find a better *hypothesis*, one that is more similar to the design solution than *DEVICE_20*. As in *DEVICE_20*, the system fails twice, when it starts from *DEVICE_13* and *DEVICE_14*. Next, the system starts from *DEVICE_88*. The system then generates a more *similar hypothesis*, *HYPOTHESIS_85*, to the design solution. However *HYPOTHESIS_85* does not satisfy all the required specifications, and its *similarity* to the design solution is 5/6. Then, the system tries to generate a more *similar* hypothesis by starting from *HYPOTHESIS_85*. Finally, the system generates two assembled machines which satisfy the given specifications. The transformation of the *component concepts* structure and the *partial function structure* to reach the design solution is illustrated in Fig. 18.

This example clearly shows the strong point of this methodology. When the machines move from the starting point to the design solution in Fig. 18, the change in the *function concept structure* is found to be slight, though the change of the component concepts is notable. For example, in the move from *DEVICE_88* to *HYPOTHESIS_85*, about half of the *component concepts* are changed, though the change in the *function concept structure* only occurs in the *function concepts* which are shaded in Fig. 18. Furthermore, in this move, the *similarity* to the design solution increases by 1/6. From this example, it can well be understood that in order to obtain the directed series of points to the design solution, particular attention must be paid to *partial function concepts*.

7. Discussion

The task of designing is simply to find design solutions which manifest the required specifications (function). For the purpose of designing per se, neither the *function decomposition process* nor the *component search process* is needed. According to Pahl and Beitz, establishing a *function structure* facilitates the discovery of solutions because it simplifies the general search and also because solutions to *sub-functions* can be elaborated separately. We have found that the nature of *decomposing functions* can also be discussed theoretically. Consider the efficiency of finding the design solutions for using the example in this study. Finding the design solutions by assembling the components, the difficulty of results in high complexity of computation, (depending on number of the combinations of the components.) Unless rules or algorithms exist for in assembling the components, this is a NP complete problem.

The approach of this study is to operate the components not individually but collectively. This ensures efficiency of calculation because the number of objects which are operated becomes smaller. However, it is difficult to operate the components collectively, using a partial function concepts, because the relation between the *total function concepts* and the *partial function concepts* cannot be adequately determined. A *metrical* relation has been found, however, and this study has developed a methodology to operate *abstract concepts* based on this very weak relation. The authors believe that this methodology to operate the *components* collectively, that is to operate *abstract concepts*, can be one approach to the *synthetic* nature in the design process. As a result, it can be theoretically concluded that the design methodologies which are distinct by explicitly dealing with the *function decomposition process* and the *component composition process* have been derived from experiential know-how in order to enhance the efficiency of the designer's search process.

REFERENCES

Pahl,G., & Beitz,W., "Engineering Design", Springer-Verlag, 1988.

Taura,T., & Yoshikawa,H., "The Generation of Design Solutions using Metric Space Methods", Intelligent CAD 3, Proceedings of The Third IFIP WG.5.2 Workshop on intelligent CAD (1989,0saka), North-Holland, 1991.

Taura,T., & Yoshikawa,H., "A Metric Space for Intelligent CAD", Intelligent ComputerAided Design, Proceedings of IFIP WG.5.2 Working Conference on intelligent CAD (1991,Columbus), North-Holland, 1992.

Tomiyarna T., & Yoshikawa,H., "Extended General Design Theory", Design Theory for CAD, Proceedings of IFIP 5.2 Working Conference (Tokyo), North-Holland, 1985.

Yoshikawa,H., "General Design Theory and a CAD System", Man-Machine Communication in CAD/CAM, Proceedings of IFIP WG 5.'/5.3 Working Conference (Tokyo), North Holland, 1982.

11

MANAGING GENERATIONAL CHANGING IN COMPUTER WORKSTATIONS

Susan Walsh Sanderson
School of Management
Rensselaer Polytechnic Institute

Abstract: This paper compares the effectiveness of Sun's product family design approach to DEC's functional module approach in managing the transition from CISC to RISC-based architectures in computer workstations. Sun's product family approach and its willingness to make an early and clean break with CISC provided for a smoother transition to new generation workstations.

1. Introduction

The mid-1980's began the period of ferment in the microcomputer and workstation industries. In the 10 years between 1981 and 1991, worldwide revenues of PCs grew from negligible proportions to over $80 billion, overtaking revenue from all other computer hardware combined. The computer industry is faced with a new competitive environment. The shift in market demand is in the direction of stand alone computing and networking solutions without a major mainframe host. Firms that grew up in the heyday of mainframe computers (IBM) and mid-sized computers (DEC) are faced with a formidable challenge. They must reorient their businesses and grow stand alone products lines such as workstations and PCs of all shape and variety. This requires a substantial shift in resources and approaches to management of innovation.

Periods of ferment and generational change often cause great difficulty for firms. Tushman and Anderson (1990) found that the "creative destruction" and new thinking that is required to launch next generation products successfully is found in both incumbent firms and new entrants. However, not all manufacturers make successful transitions to next generation products. Why are some firms better able to make this transition than others?

Although new product innovation has been the subject of academic research for decades, successful approaches to design management are not well understood. Most literature on innovation and design management focuses on individual projects and has very little to say about managing product development over time. To fill this gap, we developed a research program that has focused on analyzing product development in firms over long periods of time in major categories of products (for example, laptop computers, portable personal stereos, camcorders, and computer workstations). The findings have been reported in a series of articles (Sanderson, 1991a and b; and Sanderson and Uzumeri 1992a and b). We have demonstrated that product families and subfamilies are highly effective in helping firms organize design efforts during the incremental phase of the product life cycle in high variety and rapidly changing industries. This approach is illustrated by Figure 1.

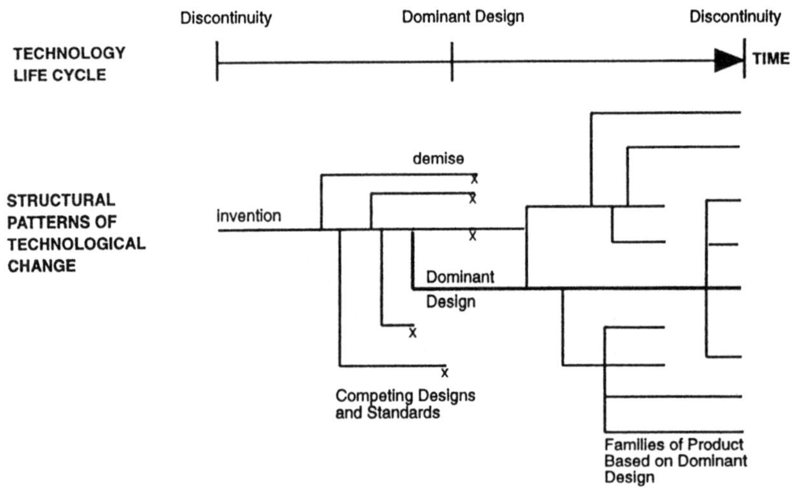

Figure 1: Product Families and the Structure of Technological Change

In those studies we found that effective management of product families involved the disciplined spinning out of models. Subfamilies were targeted to different markets and niches with clear design goals that focus on performance features such as small size or low cost. In the most innovative firms, industrial designers expanded model variety by building on and enhancing the engineering developed core platforms. Industrial designers also helped create aesthetically and ergonomically pleasing models.

We hypothesize that firms that manage product families may be faster and better able to react to generational change than firms that have a less focused design management approach. It is suggested that a product family focus provides for a greater and more effective concentration of design resources leading to faster and more efficient model development during periods of incremental change. A more focused approach can also help in making a transition from one technological trajectory to another as necessary during an era of ferment and generational change.

To better understand the difficulties and challenges associated with making generational changes we examined the shift from CISC to RISC-based computer architectures in the computer workstation industry. This transition in computer workstations is presently underway and provides an extraordinarily rich opportunity for first-hand observation. We focus on the experiences of two major incumbent firms in the computer workstation industry, DEC and Sun.

2. Computer Workstations

Since the first workstation appeared on the market in 1981, workstations have gone through several changes timed to take advantage of improvements in chip performance and price over the past decade. Successful firms in the industry have managed a series of incremental changes and are now confronted by a major generational change.

In the mid 1980s RISC-based approaches to computer design were emerging as an alternative to CISC architectures. Although computer designers such as Seymour Cray had been experimenting with this new approach to computer design for some time and Cray's machines were the fastest available for a number of years, other commercial machines were just beginning to be available.

A small group of researchers at the University of California at Berkeley headed by Patterson developed the design concepts for a RISC-based computer, the RISC I. The machine developed by Patterson and his colleagues ran with a 1.5 MHz clock speed and executed one instruction in 2 msec. IBM had RISC design projects since 1985, when John Cocke and George Radin began the 801 project that was thought to have led to the development of the RT Personal Computer. Hewlett-Packard and others began to introduce RISC machines in the late 1980s.

The commercial applications of RISC designs picked up substantially in the late 1980's when Sun, as well as others, began their own development efforts. Influenced by this earlier work, Sun put together a design team to address the future needs of the educational market.

Although improvements in the price and performance of CISC-based machines had been steady and predictable, by the mid 1980s an acrimonious debate was stirring among computer designers and proponents of the new RISC-based designs and the dominant CISC-based computer designs. Were the new RISC designs superior to CISC? Were improvements in speed sufficient to make it worth while to shift from CISC to RISC designs? How would customers migrate to applications tailored to previous designs? How much responsibility should they bear in providing for the migration of their customers to the new technology? Would the new business generated by the new product make up for the loss of the old? How quickly?

3. Design and Manufacturing Strategies of the Key Players

Although differentiation and layering in hardware, software, and networking technologies provide the basis for competitive advantage in computer workstations, in this paper we focus only on hardware design and manufacturing. In order for a firm to be successful it must be able to manage all of these elements in concert and develop and carry out a coherent and effective strategy for competing.

Most current and emerging design technologies depend on a model of product structure to achieve their effectiveness. Similarly, most design management processes are rationalized around the tasks which product structure dictates. Product architecture, modularity, family resemblance, degree of standardization, complexity and rate of change all impact the choice of appropriate design tools and the management of their implementation and use.

The workstation market grew about 60% a year between 1986 and 1989. Sun grew with the market and captured 24% of a much larger market by 1989. In this period, DEC's share also grew dramatically to 20% of the workstation market in 1989. Apollo was acquired by Hewlett-Packard and the Hewlett-Packard/Apollo combination led the industry in market share even though they lost approximately 12% combined share from the 1986 level. By 1990, there was a dramatic shift in market share. Sun's market share grew to approximately 38% of worldwide market share while Hewlett-Packard's share declined to 20% and DEC's share dropped to 16%.

After 35 years of never having a single unprofitable quarter, Digital had three consecutive losing quarters. DEC announced a $2 billion loss for it fiscal fourth quarter, which ended June 30, 1992. The loss included a $200 million operating loss and up to $1.7 billion in restructuring and accounting charges.[1] In addition layoffs of up to 15,000 are expected by the end of the 1992 fiscal year.

Writing in the *New York Times* on the occasion of the retirement of Ken Olsen, founder and former chairman of Digital Equipment Corporation, John Markoff commented:

---

[1] Digital Equipment Founder is Stepping Down, New York Times, Friday, July 17, 1992, Pg. D1.

"During the mid-1980's, Digital developed tremendous momentum around a unified business model that emphasized a single computing design from the bottom to the top of its computing lineup. Called VAX, it became one of the preeminent computers of all time. But VAX wore out as a technology in the late 1980's in the face of a new generation of microprocessor chips based on a design called reduced instruction set computing, or RISC. Even the chief designer of the VAX, Gordon Bell, warned Mr. Olsen in 1988, `Walk--don't run--away for the Vax as fast as possible.' For years, nobody listened to Mr. Bell."[2]

DEC and Sun have pursued different strategies for managing their workstations and underlying design and manufacturing tools and the transition to RISC-based architectures appears to have been far less traumatic for Sun than for DEC. Sun's performance has been steadier during this period of ferment than has DEC's. We suggest that a major reason for this greater success has been that Sun's product family oriented design teams were better able to manage both incremental and generational change. DEC, by contrast, organized its development efforts at the module level, independent of the product families for which the modules were intended. They have emphasized integration with previous designs and a reluctance to invest in radically new technologies that would cause compatibility difficulties for its installed customer base.

3.1. Digital's Workstation Strategy. Workstations were eating into the mini-computer market, the segment of the computer industry that DEC pioneered and which was responsible for its outstanding growth for over three decades. DEC entered the workstation market in 1985 with the introduction of the VAX station I followed by VAXstation II and VAX station II/GPX in 1986. Despite its late entry, in the first two years DEC was able to capture 9% of the workstation market.

Digital's basic strategy in computers has been to offer solutions using a variety of different hardware platforms, operating systems architectures, software products and support services to build a complete distributed computing system. DEC has expanded its line of products to support both RISC-based and VAX-based processors, both VMS and UNIX operating systems, and multiple bus structures and networking solutions.

---

[2] John Markoff, Stalled in the Past, New York Times, Friday, July 17, 1992, pp.D1-D2.

Digital's focus is on interoperability and compatibility of RISC, VAX, and multivendor environments. This interoperability is provided through the use of common user interface, common tools, common applications, and common networking, and DEC supports a wide range of systems and graphics standards and software tools. In supporting both VAX and RISC, VMS and UNIX, DEC believes that it is offering a key choice for new customers as well as supporting its large installed customer base.

Historically, DEC has viewed the workstation market as a necessary component in its efforts to sell integrated solutions to customers. From this point of view, it needed to be competitive but not necessarily leading edge in pure hardware price or performance. It needed to have workstations that were supported by its networking and software applications, but did not necessarily outperform the competition in any one engineering market.

This approach was consistent with its approach to modular design and production at the components level. DEC workstations were driven by innovations outside of the DEC workstation product development process itself.

DEC marketed two workstation product families, one based on the VAX computer processor and architecture, and the other based on newer RISC technologies purchased from outside of the company (see Figure 2). Another product family based on an internally developed RISC chip called Alpha is planned and will replace the current DEC stations based on the MIPS chips. The VAX based systems, called the VAXstations, support the architecture, bus structure, and VMS operating system which is common to the rest of the DEC line of computers, and has been a major part of its product line for the last fifteen years. In utilizing the VAX processor chips and the identical bus structure and operating system, the VAXstations are compatible with the DEC networking products and distributed software products used in other major DEC installations. This compatibility has been the major selling point of this workstation product line.

The second line of workstations is the DECstations, which were based on the RISC processor chip purchased from MIPS Corporation. These DEC RISC-based workstations were more compatible in performance with the competitive workstations marketed by Sun, IBM, and Hewlett-Packard. However, they were not nearly as compatible with the current or near term DEC integrated installations. They do not as easily support the applications or networking software which has been the staple of the DEC product line. The DECstation line utilizes a version of the UNIX operating system and forces difficult decisions regarding integration into a VAX-based network and distributed computing environment.

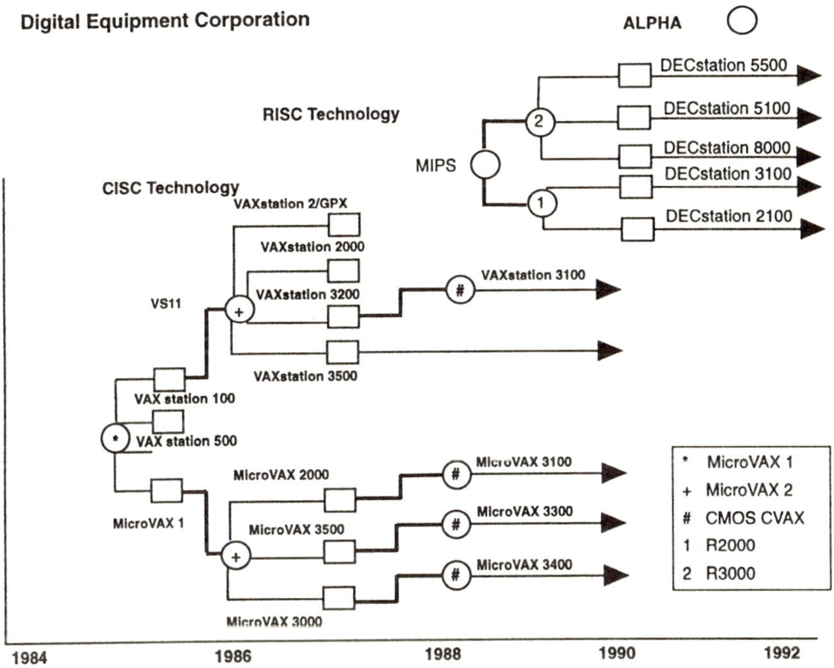

Figure 2: DEC's Workstation Products

Historically, DEC's product development strategy in the workstation market has involved the design and manufacture of modules as a process somewhat independent of the development of product lines and product families. Following this approach, DEC has been using design automation and group technology tools at the module level independently of the product line or product family for which the modules are intended. This approach is consistent with the strategy of building communications links and design constraints rather than new teams oriented around final products.

This strategy is also consistent with DEC's philosophy of relating to the sales, service and maintenance of overall integrated systems rather than targeting product lines to particular niche markets such as the workstation market. By keeping modules constant in an effort to support a variety of computing technologies and networks, they hoped to maximize the compatibility among their products and encourage customers to buy large integrated systems with supporting Mainframe computers. Maximizing or optimizing the producability and cost of the individual modules may be expected to impact the range of integrated computing technologies even though it does not optimize the marginal cost, profits, or reliability of individual product lines.

A further indication of the degree to which DEC saw the workstation market as an integrated part of its other businesses in the way in which workstations were configured and sold. Until recently workstations were custom configured by DEC salesmen out of standard modules. DEC catalogues did not give price breakdowns for variations of workstations. The subtle shift that antecede the introduction of the RISC-based workstations and the greater focus on workstations as stand alone products are revealed in the shift in price strategy to advertise prices for stand alone machines.

DEC's set for itself a very difficult strategy for competing in workstations. First, the technologies are advancing so quickly that the workstation product lines have been forced to diverge into two major product families in order to meet market needs as well as maintain the established compatibility with other DEC product lines. This loss of focus and divergence does not lend itself easily to the fully modular design and production described earlier. Second, performance of the new workstations is sufficient enough that many applications needs can be met by a stand alone network of DECstations or similar RISC-based workstations with their own server without incorporating a major Mainframe host such as the VAX computer line.

DEC's approach was to develop technology based at the level of the functional module rather than product based. The approach assumes the evolution of a systems architecture which will be sufficient to define appropriate product lines at some time in the future. The orientation of the design/manufacturing process towards the functional module reinforces the strategy for evolution of compatible and integrable products. This approach has been very successful in establishing a large base of installed computing for engineering, business, and scientific applications.

The problem that DEC is now facing is how the *evolutionary* systems-based approach can accommodate major changes in component technologies, such as those which occurred in the advent of RISC-based workstations and the UNIX operating system. The DEC strategy resulted in a shift in development change (to the RISC-based DECstation) which proceeds in parallel with the *evolutionary* VAXstation development. The carry over of functional modules becomes important to the design and manufacturing of these new products, but is often difficult to maintain in view of new functional requirements. The dual development effort necessary to deliver the continuity of the systems-driven product line and the stand-alone product family are difficult to manage at the same time.

Customer Migration: There is already evidence that DEC's migration from the VAX architecture to Alpha will be difficult and will consume considerable resources. DEC has begun the process by developing a series of tools to end users to facilitate the migration process for VMS and Ultrix software to the next generation of Alpha systems.[3] While the migration process for large, complex applications will be difficult, the dilemma over whether to provide a migration path to Alpha for all user is even greater. In an effort to save development resources. DEC decided to halt OSF/1 development of future DECstations but the internal and external uproar caused by the decision to orphan DECstation users caused DEC to reverse its decision.[4]

In the DEC marketplace, the demand for integration and compatibility has had a higher priority than the fast product cycle maximum performance required in other commercial markets, particularly the individual engineering workstation. DEC has managed the design and manufacturing for the VAX product family around the evolutionary systems market, and has evolved its design and manufacturing tools and management structure around that model. As revolutionary rather than evolutionary changes are required to keep up a competitive performance of their overall systems, will DEC be able to respond or will it be necessary to develop a more product oriented approach?

This question has particular relevance with DEC's announcement of its intentions to build a new family of workstations around Alpha chips. The Alpha chip, the first of a series of RISC chips, processes 64 bits of information at a time, twice as many as most current computers.[5] Alpha is able to process as many as 400 million instructions a second and estimated to significantly exceed the current 28 million instruction speed of Sun Microsystems Inc.'s Sparc chip. Sun and Hewlett-Packard have also recently announced new RISC chips that were significant improvements over previous chips.[6]

DEC's announced strategy in managing the new Alpha line of workstations is to be the price and performance leader. But it is not clear whether there will be a significant departure for current methods of managing product development described above. Will the Alpha architecture be sufficient to carry through Digital's traditional line of products without adopting a more

---

[3] Elinnda-Carol Ballou, "DEC to unleash tools for Alpha end users, Computerworld, June 22, 1992, pp. 14.

[4] ibid, p. 14.

[5] Glenn Rifkin, Digital's Chip of the Next Century, New York Times, February 25, 1992, pp.D1 & D5.

[6] ibid.

disciplined approach to managing individual product families such as computer workstations?

3.2. Sun's Workstation Strategy. Sun Microsystems introduced its first computer workstations based on the Motorola 68010 chip in 1983. Prior to 1989, Sun followed a systematic evolution of its workstation family based on Motorola processors. During the first six years of its existence Sun grew from a small start-up to a $1 billion business, growing the following year by 70% with sales of 1.7 billion. As figure 3 demonstrates, Sun has been very systematic in spinning out its models during the incremental phase of development.

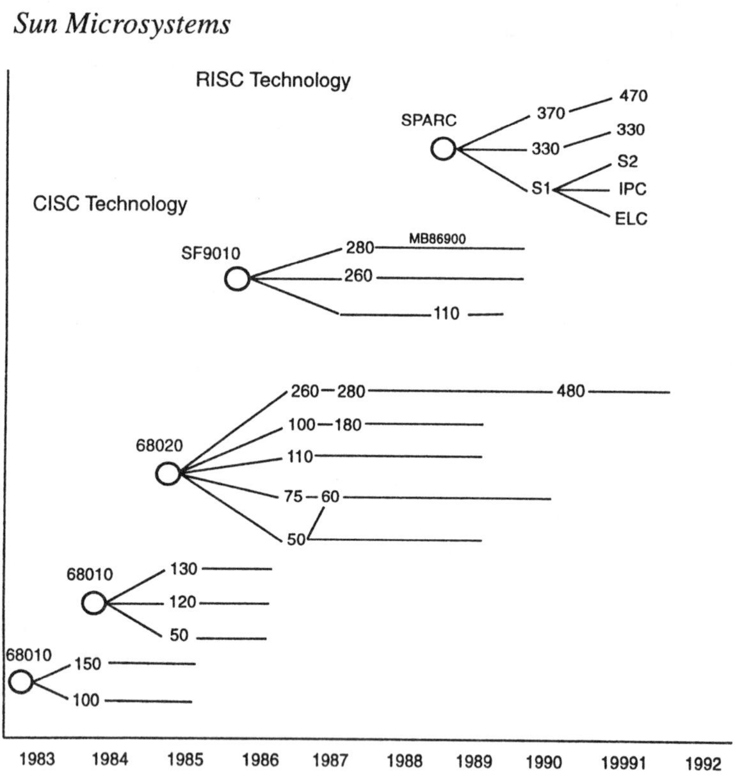

Figure 3: Sun's Workstation Families

Sun began the development of its first RISC-based workstation in the late 1980s. Originally designed with the educational market in mind, RISC was a foray into the unknown. In 1989, Sun developed a new family of workstations built around RISC architecture and introduced the Sparc chip, a microprocessor of its own design. Sun does not produce the Sparc chip itself but outsources its production. It also licenses the design to chip manufacturers such as Fujitsu, Philips, Texas Instruments, and Toshiba and encourages other companies to design and build Sparc clones (Rappaport and Halevi, 1991). In licensing the Sparc chip, Sun hopes to improve the performance and reduce the cost of the chip. In making the chip available to other manufacturers, Sun assures that there will be other hardware and software compatible with its own designs. Sparc architectures now account for 60% of the workstation industry's RISC unit volume and independent software vendors have developed more than 2,100 applications for Sparc hardware--more than twice the number available for workstations designed by DEC and IBM (Rappaport and Halevi, 1991).

Sun made a significant advance in market share in 1990 with the introduction of its Sparc workstation platform. Sun presently has almost 40 percent of the worldwide workstation market, pulling ahead of its nearest rivals, Hewlett-Packard and DEC.

In 1989 and 1990, models from both CISC and RISC families were sold. Sun made the dramatic decision in 1991 to discontinue its older workstation platform based on the Motorola chips and focus all of its efforts on the new RISC-based platform. Unlike the DEC approach that is trying to support both VAX-based and RISC-based families, once Sun established its new RISC-based platform it made a clean break with the past by dropping its older platforms. Now Sun is managing the new RISC-based platform using an evolutionary strategy as it systematically develops models of the Sparc station family.

4. Comparing the DEC and Sun Strategies

Figure 4 shows a comparison between DEC and Sun's management of its CISC technologies and investment in new RISC architectures. As we can see, Sun invested in the full development of a RISC line of chips earlier than DEC. DEC dragged its heels in making an investment in RISC and then did so in a half-hearted way. The technology development, namely development of a new class of computer chip, began in earnest at DEC with the Alpha chip. By contrast, Sun's early investment set the stage for the development of a new family of workstations.

DEC's initial approach was to minimize the change in its product offerings by purchasing new RISC chips from external suppliers rather than invest in the costly and difficult development of a new architecture that would compete directly with its traditional VAX family of products. That strategy proved to be only a stop gap measure and eventually led to further erosion of its traditional marketplace.

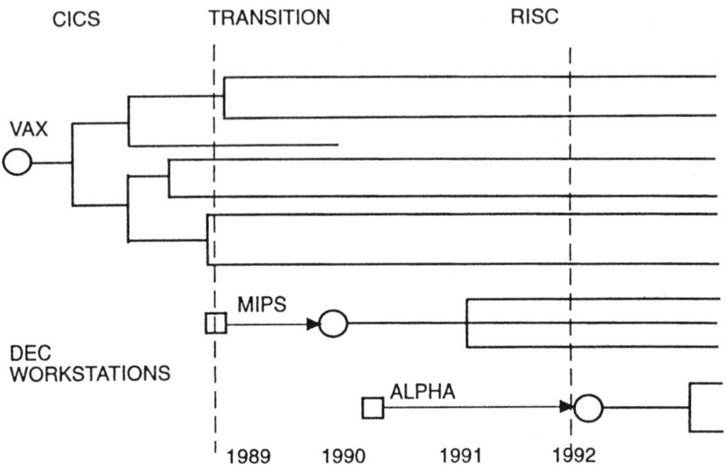

Figure 4 Workstations Development at DEC and Sun

DEC had built its empire on the VAX architecture and could not find a painless migration either for itself or its customer. It waited too long to make the necessary transition.

One of the reasons that DEC may have reacted so slowly was its view of its mission. As one of the pioneers of minicomputers and later midrange computers, DEC saw itself as filling a wide mission with a large established customer base to which it owed its success. The outstanding success of the VAX family of products and its large installed base made it difficult to pursue technology directions that might make obsolete the equipment in which loyal DEC customers had already invested. DEC's commitment to providing a migration path for all of its users made it very difficult to make the radical break that RISC required. When DEC decided to invest in RISC it did so in a half-hearted way. First it purchased chips from MIPs and did much of the design work in California away from the center of most of its new design activities.

When it was clear that the half-hearted investments would not be enough to offer truly competitive workstations to lure new customers and compatibility with older machines was not enough to keep the older customers, DEC began a serious investment in RISC. The new Alpha chips hold much promise to provide the scaleable architecture necessary to form a strong framework for DEC's next generations of products. If it is successful in building products around the Alpha chip as it was around the VAX architecture it will be a formidable player in the computer industry for decades to come.

5. Conclusions

The transition from one generation of workstation to another is not without difficulty. The shift in architectures, operating systems, and network protocols required by major changes in technology represents a significant challenge for firms. In this paper we have described the management of incremental and generational change at DEC and Sun. We have demonstrated that effective management of product families can be a useful organizing principle in managing both incremental and generational change in computer workstations. Sun was effectively managing a single product family in hardware, the computer workstation. It already had a clear product family focus and when the time came for the transition to RISC Sun made an early an focused development effort. It was willing to make an abrupt break with the past and provided customers support to help them migrate their software.

By contrast, Digital never h a d a product family approach for its workstation market. It saw workstations as extensions of its other product lines. This confusion made it much more difficult to commit to change and harder to implement the change once the decision is made to develop a new technical base.

DEC's approach to the development of its design tools reflects its strategy to build communications links and design constraints rather than new teams oriented around final products. Sun's focus on end product families has led to an emphasis on a different design approach and the systematic development of successive models within the context of a single product family.

The management of parallel product families, e.g. DEC's VAX and RISC-based families, is a difficult balancing act and requires the commitment of significant design and manufacturing resources. The problems of managing parallel product families are even more difficult in the context of a fast-paced technology-driven environment.

Firms face a significant challenge in fast-paced industries with high rates of technological change. On the one hand they must be consistent and systematic in developing the variety of models at a cost and performance that the market requires, while at the same time they must invest in the new technologies that will obsolete their products. Sun elected to discontinue its older generations of workstations, sacrificing compatibility with previous generations for performance. DEC is trying to support two different workstation platforms as well as multiple buses and operating systems. These alternative strategies for a given product are based on the wider scope of development and marketing decisions which govern the company-wide business plans.

REFERENCES

Clausing, Donald, Flexible Product Development, Massachusetts Institute of Technology, unpublished research report, no date.

Digital Equipment Corporation, VMS Workstations Product Strategy, internal report, January 8, 1991 & February, 1991.

Digital Equipment Corporation, Workstation Strategies & Directions, internal document, no date.

Gomory, Ralph, From the `Ladder of Science' to the Product Development Cycle, Harvard Business Review, November-December 1989.

Hall, Mark and John Barry, Sunburst: The Ascent of Sun Microsystems, Contemporary Books, Chicago, 1990.

Harris, Marilyn, et. al., "Can Apollo and Sun Hang On to Stardom?" Business Weak, January 14, 1985, 126J.

Rappaport, A.S. and S. Halevi, The Computerless Computer Company, Harvard Business Review, July-August, 1991, p.69-80.

Sanderson, Susan W., Strategies for New Product Design and Product Renewal, Academy of Management Best Papers Proceedings 1989, p. 296-300.

Sanderson, Susan Walsh and Vic Uzumeri, Industrial Design: The Leading Edge in Product Development for World Markets, to appear in Design Management, forthcoming.

Sanderson, Susan Walsh, Design for Manufacture in an Environment of Continuous Change, to appear in Gerald Sussman, ed. Design for Manufacture, Oxford University Press, 1991.

Sanderson, Susan Walsh, Cost Models for Evaluating Virtual Design Strategies in Multicycle Product Families, to appear in Journal of Engineering and Technology Management, 1992.

Uzumeri, Vic, The Challenge of Product Variety and Change for Manufacturing, Rensselaer Polytechnic Institute, unpublished doctoral dissertation, August, 1991.

12

FUNDAMENTALS OF PRODUCT MODULARITY

Karl Ulrich
Massachusetts Institute of Technology
Sloan School of Management, Room E53-390
Cambridge, Massachusetts 02139
USA

Acknowledgments: This research was supported by the National Science Foundation under contract number DDM-8914181 and by the MIT Leaders for Manufacturing Program, a partnership between MIT and 11 major corporations. Many of the ideas in the paper were developed jointly with Karen Tung during a collaborative research effort with the Square D Company. I am grateful to David Ellison, Frank Gillett, Joe Pine, Henry Stoll, Mohan Tatikonda, and Daniel Whitney for their contributions to the ideas in this paper. This paper is a condensed version of a paper with the same title (co-authored with Karen Tung) in *Proceedings of the 1991 ASME Winter Annual Meeting Symposium on Issues in Design/Manufacturing Integration* (ASME DE-Vol. 39).

Abstract: Modularity arises from the division of a product into independent components. This independence allows a firm to standardize components and to create product variety. Product modularity is frequently stated as a goal of good design practice, yet has not been as carefully studied as some other current design practices such as design-for-assembly. As firms strive to rationalize their product lines and to provide ever increasing variety at lower cost, modularity is becoming a focus of attention. The goal of this paper is to define modularity, to explore some of the benefits and costs of modularity, and to propose a set of research questions. The paper is the result of preliminary research and is more a starting point than a set of refined conclusions.

1. Introduction

Product modularity is frequently stated as a goal of good design practice (see, for example, Pahl and Beitz 1984 chapter 7 and Suh 1991). In the cases of Nippondenso automotive panel meters (Nevins and Whitney 1989), the

Xerox Memorywriter typewriter (Jacobson 1986), and IBM printers (Society of Manufacturing Engineers 1990), modularity is credited with many of the products' successes, particularly low production costs. Yet modularity has not been as carefully studied as some other current design practices such as design-for-assembly. As firms strive to rationalize their product lines and to provide ever increasing variety at lower cost, modularity is becoming a focus of attention.

The goal of this paper is to define modularity, to explore some of the benefits and costs of modularity, and to propose a set of related research questions. We focus primarily on modularity in the domain of discrete physical products as opposed to chemical products, electronic circuits, or software products, although many of our ideas appear to be valid in these domains as well.

2. What is Modularity?

The term *modularity* is used informally in many ways. In the design of complex engineered systems (e.g. space stations) the term often refers to the use of *independent* units. In architecture the term usually refers to the construction of a building from many instances of *standardized* components. In manufacturing the term often refers to the use of *interchangeable* units to create product variants. These definitions are highly interrelated. In fact, we will argue that independence of components is the property of a design that allows standardization and interchangability.

Modularity arises from the way a product is physically divided into components. For our purposes, modularity is a *relative* property; products cannot be classified as either modular or not, but rather exhibit more or less modularity in design. We view modularity as depending on two characteristics of the design:

1. Similarity between the *physical* and *functional* architecture of the design.
2. Minimization of incidental interactions among physical components.

Physical and functional similarity. A product[1] can be described functionally by a collection of functional elements linked together by exchanges of signals, material, and power. This kind of description is frequently called a *schematic* description (Ulrich and Seering 1989). For example, a conventional automotive power system can be described by a collection of functional elements that might include: an element for converting a fuel and air flow to rotary mechanical power, heat, and products of combustion (an engine); an element for converting rotary mechanical power to electrical power (an

---

[1] We use the term *product* to refer both to self-contained artifacts provided to end users and to systems and subsystems that are the focus of a design effort.

alternator); an element for converting rotary mechanical power at one speed and torque to rotary mechanical power at another speed and torque (a transmission); and an element for converting rotary mechanical power to fluid flow (a water pump). The degree to which this functional description is mirrored by the physical architecture of the product contributes to design modularity. For example, if the engine and transmission of the power system were physically embodied in the same physical component (as is done in many motorcycles) then the design would be less modular than if the engine and transmission were separable physical components (as is the case for most automobiles).

Minimization of incidental interactions. The second characteristic of modularity is the degree to which the interactions between physical components are confined to those critical to the function of the product. All of the interactions specified in the functional description of the product are critical; every exchange of power, signal, or material contributes to the device function. Even if the product is physically divided into components corresponding in a one-to-one way to functional elements, in reality there may still be many other incidental interactions between components. For example, in most automobiles the heat of the engine will influence the seal on the water pump. This interaction is entirely incidental to the function of the engine and the water pump. As these incidental interactions are eliminated the design becomes more modular.[2]

To illustrate this definition and the differing degrees to which designs can be modular, consider three ways an automobile engine and alternator could be configured (figure 1). The first alternative is to use the crank shaft of the engine as the armature shaft of the alternator and to mount the field coils directly in the engine block. The second alternative is to create a separate component, containing the armature and field coils of the alternator, that mounts concentrically to the front of the engine block. The third alternative is an alternator assembly that mounts to the side of the engine block and is coupled to the crank shaft by a v-belt transmission. The key distinction among the three alternatives is the physical division of the system and the nature of the interactions between the alternator and the engine. In the first configuration, the engine and alternator are physically integrated and interact thermally, structurally, kinematically, and spatially. Some of these interactions are critical to the function of the system and some of them are incidental. In the second

---

[2] In software engineering these incidental interactions are called *side effects*.

configuration the engine and alternator have been physically separated, although there are still many incidental interactions between them. In the third configuration, in addition to physical division of the functional elements, the interactions between the engine and alternator are largely reduced to the functionally essential exchange of rotary mechanical power. By our definition, the three designs are shown in order of increasing modularity left to right.

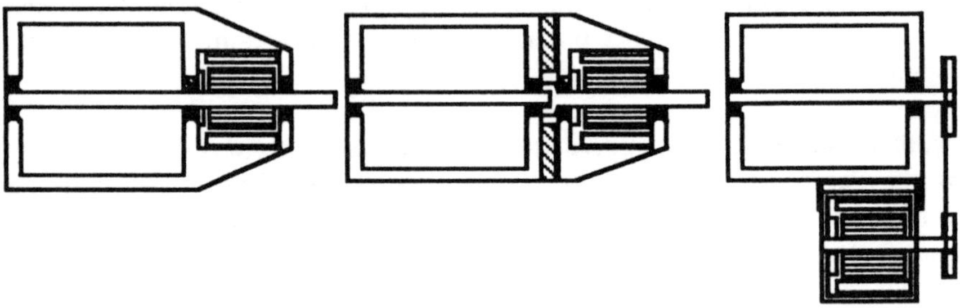

Figure 1: Increasingly modular approaches to designing engines and alternators.

A completely modular design embodies a one-to-one correspondence between each functional element of the design and a single physical component, in which every interaction between components is critical to the function of the system. To our knowledge, no products achieve complete modularity, although some electronic products exhibit relatively high modularity. Consider, for example, a product consisting of a computer and a printer. The computer and printer as separate components reflect the functional architecture of the product. And to a great extent the only interactions between the computer and printer components are essential to the function of the system— the exchange of digital signals across a cable. By our definition, the design is highly modular. There are, however, some incidental interactions. If, for example, both devices are placed in a small closet, the heat from the printer may cause the computer to fail. There is also a spatial constraint on physical proximity (an incidental interaction) imposed by the length of the cable.

3. Potential Benefits of Modularity

In this section, we articulate a set of potential benefits of modularity that seem to underlie much of industrial practice. Some of the benefits relate to design and production economies, some relate to customer responsiveness, and others relate to the organization and operation of the design and production systems.

• *Component economies of scale.* Modularity allows the same component to be used across product variants and across product lines. This standardization is possible because in a modular design the function of a component is well defined and physically separable, and because the incidental and idiosyncratic interactions between a component and the rest of the product have been minimized. When a component is standardized and used in several product lines, its production volume is higher than if it were designed and used in only one particular product line. Standardization allows development resources and capital expense to be amortized across a large number of units and allows the exploitation of higher-volume, more-efficient production technology in component manufacturing. These economies may be enjoyed directly by the manufacturer or may be manifest by the relatively low cost of a standard component supplied by a vendor. An example of this benefit is the use of a standard motor in the drive system of a computer printer. By dividing the printer into components that allow for a standard motor interface, the manufacturer may reduce the cost of electromechanical actuation from hundreds of dollars per unit to a few dollars per unit. In addition to lower costs, standardization takes advantage of the increased reliability and technical performance that often evolve in a high-volume standard component.

• *Product change.* Modularity benefits the ease with which a product can be changed. Desired rates of change vary among the different components in a product. These differences can arise from customer preferences or technological advances. For example, consumer electronics enclosures must change more rapidly than the underlying electronics. Automobile body styles change more rapidly than power trains. Memory chip technology changes more quickly than power supply technology. By dividing a product into components and interfaces with different desired rates of change, a manufacturer can accommodate necessary change without disrupting the design of the entire product (Hagel 1988, Sanderson and Uzumeri 1990). Another form of product change is the reuse of elements of an existing design in a subsequent design. If an existing product has well-defined interfaces and components, then the design of the next product can borrow heavily from the components of the previous product (Rothwell 1990). Forms of product change of direct importance to the end user are product upgrade and extension. These benefits are more easily achieved in modular designs. An example of such a benefit is the upgrade of the logic board in a personal computer.

• *Product variety.* One of the most common motives for modularity is that it allows a large variety of end products to be constructed from a much smaller set of different components. The variety available in modular designs arises from the ability to use one of several alternative component options to implement a functional element of the design, or to combine different sets of components to achieve different overall device functions. This substitution and combination is possible because of the one-to-one mapping of component boundaries to the boundaries of functional elements and because the interfaces between components can be well defined. The alternative would be to design a distinct product for each desired application. The appropriate use of modularity can be dramatically more efficient than this alternative. For example, IBM was able to create seven different printer motors by varying only the face plates, gears, and end caps of the motor while leaving 25 parts unchanged (Society of Manufacturing Engineers 1990). These efficiencies reduce the complexity and resource requirements for the design and production system of the manufacturer (Nevins and Whitney 1989).

- *Flexibility in use.* In addition to allowing a manufacturer to create variety, modularity may offer the ability to vary the product in use. For example, many passenger vans exhibit modularity in their seating systems, allowing separation and removal of the seats when cargo space is desired. Many single lens reflex 35mm cameras allow different lenses to be interchanged.
- *Order lead-time.* Whenever products must be made to order because of customer demands for customization, modularity allows the order lead time to be shorter than the lead time for the individual components that make up the finished goods. One type of situation in which this is true is when a make-to-order or custom component is married with an otherwise standard product. In this situation, the bulk of the standard product can be inventoried and then the customization can be focused on the differentiating component. A similar strategy can be adopted when product variety is achieved through the *combination* of standard components. The standard components can be inventoried and then the product can be configured to order. This strategy has been called the mushroom approach because the product variety "mushrooms" at or near final assembly (Mather 1987).
- *Decoupling of tasks.* Dividing a product into components requires the definition of interfaces. These interfaces enable design tasks to be decoupled and allow production tasks to be decoupled. This decoupling results in reduced task complexity and in the ability to complete tasks in parallel. Once an interface is defined in a modular design, the component's interactions with the product have been largely confined to that interface. Although there are inevitably interactions that occur beyond the defined interfaces (e.g. the heat from an engine *will* affect the alternator even if the alternator is a completely separate component), this confinement greatly simplifies the interactions that the designers must anticipate. The benefits of decoupling extend to the production of the components as well. Independent components can be

produced and tested separately. For both design and production, the division of a product into independent components also allows the parallel completion of tasks. These benefits of decoupling are credited with much of the success of Japanese auto manufacturers in reducing product development lead times (Clark and Fujimoto 1991).

• *Design and production focus.* Dividing a product into independent components allows design and production activities to be specialized and focused. This focus may manifest itself in terms of special facilities. For example, clean-room environments can be confined to disk drive manufacturing and do not need to extend to the entire computer production system. Another instance of production system focus is the development of components around the capability of work cells in the factory (Shirley 1990). This focus could be called an *economy of specialization* (von Hippel 1990).

• *Component verification and testing.* Because the components in a modular design correspond to particular functional elements, the function of the component is well defined and a functional test should be possible. Because the interactions between components in a modular design have been confined to those that are critical to the function of the product, the interface between the component and the rest of the product can be relatively easily simulated.

• *Differential consumption.* Similar to the benefit of product change is the benefit of differential consumption. This motive underlies the design of vacuum cleaner bags, razor blades, and film. Because these elements of the product are consumed faster than the rest of the product, they are appropriate choices for components in a modular design. The alternative strategy is to drastically reduce the cost of the non-consumable portion of the product as is done with single-use razors and cameras.

• *Ease of product diagnosis, maintenance, and repair.* These benefits are accrued after the product is put into service. In some cases, components have become replacements for test instruments; replace a component rather than diagnose a specific fault. Exchanging components is facilitated by modularity, thus improving the speed with which a product can be repaired.

4. Potential Costs of Modularity

In this section we outline some of the potential costs associated with modularity. The first three costs arise directly from modularity. The remaining costs arise from the use of modularity to facilitate component standardization and product variety.

• *Static product architecture.* Modularity assumes a particular product's functional and physical architecture. This assumed architecture can be an obstacle to architectural innovation, given that it often defines the organization of the firm and the knowledge that is embodied in the firm's product delivery system. This problem has been identified by Henderson and Clark (1990) in the photolithography industry and may in fact be of concern in many other industries as well. The importance of the product architecture in organizing and managing product development has also been underlined by other researchers (von Hippel 1990, Clark 1985).

- *Performance optimization.* Given unlimited engineering resources, the technical performance of any particular product can always be improved by reducing modularity. This performance improvement is often in terms of reduced size and mass; modular designs can contain redundant physical structure and do not exploit as much function sharing (the implementation of several functional requirements by a single physical element) as is possible (Ulrich and Seering 1990). Eliminating interfaces, can remove redundant physical elements from a design. For example, if the alternator and engine were not designed as separate components with a well defined interface, much of the physical structure associated with the alternator could be eliminated. The crank shaft could also be the alternator shaft and possibly eliminate shaft material, a belt drive system, and a set of bearings. By enclosing the alternator in the engine block, the alternator housing as well as the support bracket could probably be eliminated.
- *Ease of reverse engineering.* Observing a modular design is analogous to reading the schematic diagram that the product engineers used in designing the product. The function of components is often obvious and the interconnections between components are clear and well defined. These factors contribute to an ease of reverse engineering of a product by competitors. The same properties that make a design easy to reuse by the original manufacturer make the design easy to copy by competitors.
- *Increased unit variable costs.* Particularly when modularity is used to exploit economies of scale from the standardization of components, the components used across several product lines may have excess capability when used in some particular applications. This is because the component must be designed for the application with the most stringent demands. For example, an electrical ground cable used for all automobiles in a product family will have to carry the current of the highest amperage electrical system in the product family. This requirement will result in excess copper in the cable of models with lower amperages. In this situation the firm in effect gives away excess capability. In some cases, this excess capability might have higher associated unit variable costs than a component designed specifically for a particular application, although the overall cost of providing the component functionality to the entire product line may be lower (Tung 1991).

• *Excessive product similarity.* If modularity is used as a way to construct products from standard components, there is a risk of not perfectly meeting the requirements of some customers. Some auto makers have suffered from this problem because of the customer perception of too much sharing of styling and systems across models. For this reason, much of the standardization of components that can be achieved with modularity is confined to hidden components, particularly in consumer goods (Pahl and Beitz 1984).

5. Research Questions

This paper describes only the first step into a complex set of research problems. We have attempted to provide some basic definitions and to create a vocabulary for discussing product modularity issues. We close the paper by posing four questions which we have begun to address as the next steps in this research: What are the economics of standardization? What is manufacturing system complexity and how does it relate to market variety and product architecture? What is a rationale and methodology for decomposing products into modules? What are the linkages among product architecture, the production system, supplier relations, and organizational structure?

5.1. Economics of standardization. Whenever there are costs and benefits surrounding a decision, there is an opportunity for decision support models to be useful. To our knowledge few if any principled methods are in use to decide how much standardization is optimal. In the engine and alternator example, a manufacturer must decide which automobile models should share an alternator. The potential cost of sharing an alternator is giving away excess alternator capability in less demanding models. The potential benefits are economies of scale and reduced design and production system complexity. We have modeled this simple case in a field study in the electrical distribution equipment industry (Tung 1991) and found that the basic trade-off is between (a) lowering material costs by increasing the number of different components to exactly match each different application and (b) lowering tooling and overhead costs by reducing the number of components that are used across the different applications. As one would expect, in cases where materials are very expensive (copper bus bar, for example), specialized components can be economically justified, while in cases where material costs are relatively low (sheet metal enclosures, for example), the number of different components should be minimized. This analysis has been made possible by the recent application of activity-based costing (ABC) methodologies by many manufacturers. The data from ABC analyses can be used to better estimate the overhead costs associated with changing the number of different components used in a product line. Our initial experience with a single product line is that the current number of components was five times greater than the model would suggest is optimal. While there has been formal theoretical research in modeling some of these decisions (Evans 1963, Shaftel 1971, Shaftel and Thompson 1977), little work has been done to include the issues that are relevant in a typical manufacturing setting.

5.2. Manufacturing complexity, market variety, and product design. An ideal situation for a manufacturer would be to be able to provide infinite product variety to the marketplace with what appears to the manufacturing system to be a single product. There are some situations that approach this ideal. For example, Lutron, a manufacturer of lighting systems, claims to never have shipped the same system twice. They have achieved this variety through clever use of modularity in their product design. Pine (1992) has thoroughly documented this and other examples of *mass customization* — a strategy which achieves product variety through design modularity and component economies of scale.

We are interested in defining manufacturing system complexity and in developing ways of describing this complexity for a particular manufacturing system. We believe that the ability to characterize the manufacturing complexity that will be incurred in creating the desired product variety with a particular product architecture is an essential first step in supporting product and process design decisions aimed at achieving mass customization.

5.3 Rationale for product decomposition. When designing a new product or redesigning an existing product, there are typically hundreds of apparently reasonable alternative ways to decompose the product into components. Where should the component interfaces be, and by implication which functional elements should be allocated to physical components? There appears to be some consensus among experienced systems engineers as to what makes a good product decomposition, although the rationale behind these opinions is often implicit and subjective. We are not hopeful that product decomposition decisions will soon yield to an algorithmic solution, but do believe that much more systematic approaches can be brought to bear on the process than is current industrial practice.

Based on preliminary thinking, several principles seem to provide powerful decomposition heuristics. For example, functional elements that will be implemented with rapidly evolving technology should be separate components to allow easy incremental product change; the engine control module is an example of a decision adhering to this heuristic. Functional elements with "portable" interactions are candidates for physical separation; digital signals are much more portable than mechanical connections and so digital electronics should be separated while mechanical systems should be physically integrated. Functional elements that must change in response to market turbulence should be implemented as separate components; enclosures should be separate from tape-transport mechanisms in VCRs. Functional elements that contribute "holistically" to customer satisfaction should be physically integrated; instrument panels in autos are one such example.

As in the economics of standardization, there has been some theoretical research on decomposing systems (Alexander 1964), but very little work in addressing the full range of issues that are relevant in industrial practice. Our vision is of decision support tools that would suggest alternative product decompositions given the functional elements embodied by the product and the market and production system context.

5.4. Linking product architecture, production, suppliers, and organizational structure. There are strategic and organizational implications of product modularity. We suspect that at a particular point in time, the best product architecture for General Motors will be different from that of Volvo. The two firms have dramatically different production systems, organizational structures, supplier relations, and internal capabilities. Volvo is not likely to embark on the development of a new integrated engine and alternator, while it may be advantageous for GM to do so. We fear that without understanding which of these contextual factors are most tightly connected to product architecture, prescriptions for better product design decomposition are likely to fail.

We have begun to address these issues by characterizing the differences in product architecture that are exhibited by different manufacturers in the automobile industry. Our initial efforts have been focused on door systems. Our approach is to create functional and physical descriptions of products through the use of what we call *product archeology*, the disassembly and description of the artifacts produced by a manufacturer. Once we have verified that there are indeed elements of difference in product architecture across different manufacturers we can begin to address the question of whether these differing decomposition strategies make sense in the context of the individual firms.

REFERENCES

Alexander, Christopher, Notes on the Synthesis of Form, Harvard University Press, Cambridge, 1964.

Clark, Kim B. and Takahiro Fujimoto, Product Development Performance, Harvard Business School Press, 1991.

Clark, Kim B., "The Interaction of Design Hierarchies and Market Concepts in Technological Evolution," Research Policy, (14) p. 235-251, 1985.

Evans, David H., "Modular Design— A Special Case in Nonlinear Programming," Operations Research, Vol. 11, p. 637-647, 1963.

Hagel III, John, "Managing Complexity," The McKinsey Quarterly, Spring 1988.

Henderson, Rebecca M. and Kim B. Clark, "Architectural Innovation: The Reconfiguration of Existing Product Technologies and the Failure of Established Firms," Administrative Science Quarterly, (35) p. 9-30, 1990.

Jacobson, Gary, and John Hillkirk, Xerox: American Samurai, MacMillan Publishing, 1986.

Mather, H., "Logistics in Manufacturing: A Way to Beat the Competition," Assembly Automation, (7) p. 175-178, 1987.

Nevins, James L. and Daniel E. Whitney, Concurrent Design of Products and Processes, McGraw-Hill 1989.

Nussbaum, Bruce and John Templeman, "Built to Last—Until it's Time to Take it Apart," Business Week, September 17, 1990, p. 102.

Pahl, Gerhard and Wolfgang Beitz (Edited by Ken Wallace), Engineering Design, The Design Council, London 1984.

Pine, B.J. II, Mass Customization: The New Frontier in Business Competition, Harvard Business School Press, 1992.

Rothwell, Roy and Paul Gardiner, "Robustness and Product Design Families," in Mark Oakley (Editor), Design Management: A Handbook of Issues and Methods, Basil Blackwell, Cambridge 1990.

Sanderson, S. and M. Uzumeri, "Product Change and Variety: The Case of the Sony Walkman," RPI mimeo 1990.

Shaftel, Tim, "A Simplex-Like Algorithm for the Continuous Modular Design Problem," Operations Research, Vol. 25, No. 5, p. 789-807, 1977.

Shaftel, Timothy L. and Gerald L. Thompson "An Integer Approach to Modular Design," Operations Research, Vol. 19, p. 130-134, 1971.

Shirley, Gordon V., "Models for Managing the Redesign and Manufacture of Product Sets," Journal of Manufacturing and Operations Management, (3) p. 85-104, 1990.

Society of Manufacturing Engineers Manufacturing Insights Videotape Series: Design for Manufacturing, 1990, SME P.O. Box 930, Dearborn, Michigan 48121.

Suh, Nam P., The Principles of Design, Oxford University Press, 1990.

Tung, Karen, "Modularity and Component Sharing as a Product Design and Manufacturing Strategy," MIT Sloan School of Management, SM Thesis 1991.

Ulrich, K.T. and W.P. Seering, "Function Sharing in Mechanical Design," Design Studies (11) p. 223-234, 1990.

Ulrich, K.T. and W.P. Seering, "Synthesis of Schematic Descriptions in Mechanical Design," Research in Engineering Design, (1) p. 3-18, 1989.

von Hippel, Eric, "Task Partitioning: An Innovation Process Variable," Research Policy, (19) p. 407-418, 1990.

13

RATIONALIZING THE DESIGN PROCESS

William S. Lovejoy
Graduate School of Business
Stanford Univesity
Stanford, California

Abstract: There is a strong sympathy between the literature on organizational design and that for product design, as one would expect. Product design is just one manifestation of the general organizational problem: how to divide up a complex of tasks into well defined subtasks, and then coordinate the divided whole to pursue a focussed objective. This paper outlines the sympathies between these two literatures, and shows how product design techniques such as conjoint analysis, quality function deployment, Taguchi methods, design for assembly, value engineering, failure modes and effects analysis, empirical guidelines, team building, and organizational initiatives address various facets of the general coordination problem.

1. Introduction

The Society of Manufacturing Engineers handbook (Stoll 1988) defines design as pulling together something new or arranging existing things in a new way to satisfy a recognized need; purposeful planning as revealed in or inferred from the relations of the parts to the whole. Implicit in this definition is the notion of purposive action, and the notion of integration, of a whole being derived from the interrelationships among its parts. The term rationalizing means making conformable to reason, or interpreting from a rational standpoint. Rationalizing the design process means adopting a mental lens through which one can see the relation of the parts to the whole in the process itself.

The problem of how to organize many diverse individuals to act coherently toward a global objective has been addressed from the perspectives of economics, organizations, and human resource management, among others. The design community has developed its own methods and insights specific to the product development process. A typical product design and launch project requires the coordination of hundreds or thousands of interdependent tasks. Humans, however, are boundedly rational, that is, they have a finite capacity for information processing. Consequently, no single individual can micro-manage the entire design process in a reasonable amount of time. This limitation necessitates a division of labor into subtasks of reduced complexity. Continued divisions of subtasks into sub-subtasks, etc. will continue until the complexity of the divided tasks is appropriate for the available design technologies and time constraints. Following a division of labor, each individual work team will have discretion over only a part of the totality of activities. Some means must then be provided to coordinate the actions of this diverse population. In summary, the fundamental organizational problem is to:

a) Divide up the complex project into manageable subtasks, and

b) Reintegrate the divided whole to account for interdependencies and address the global objectives.

Reintegration activities will result in overhead expense (involving negotiation, experimentation, or managerial initiatives, see below). If the tasks can be divided up so that less coordinative effort is necessary, then the firm can operate with lower overhead. The ideal organization will have each subtask in the divided whole executed independently and simultaneously without coordinative effort. This is seldom possible, but the closer one can get to this ideal of completely autonomous tasks, the lower the overhead costs of coordination will be. Consequently, the overhead minimizing division of labor will be one that minimizes the interdependencies between subtasks. Stated differently, strongly interdependent activities should be put into the same subtask.

1.1. Integration techniques. Three integration techniques have found voice in the literature: market-based, constraint-based, and values-based integration. These have been called different things by different research communities. For example, they have been referred to as the market, technical or scientific, and commitment or mutuality models of human resource management (*cf* Lawrence 1985). In the organizational cultures literature they have been referred to as markets, bureaucracies, and clans (*cf* Ouchi 1980,1981). The transactions economics literature has identified markets and hierarchies (*cf* Williamson 1975), and has not addressed values-based models. A formalized version of the problem of division and reintegration appears in decomposition techniques for mathematical programs in the management science literature, where price-based and constraint-based techniques are analogous to their namesakes here (*cf* Moore 1979). Notions of organic and mechanistic organizations (*cf* Burns and Stalker 1961) are similar to attributes that attend market-based and constraint-based integration. Briefly, these integration techniques are characterized as follows:

Market-based integration: Market-based integration relies on barter among individuals. Parties with individual agendas meet and discuss tradeoffs that they are willing to accept. In a design project, representatives of product design and process engineering may meet and negotiate design attributes. This is the sort of team effort, simultaneous, concurrent, or decentralized decision processing that has received a lot of attention in the design literature. The divided tasks are left to their own devices to find tradeoffs among the variables that each controls. The archetype is a perfectly competitive market as idealized by Adam Smith.

Necessary for negotiation and barter are two things:

a) There must be discretionary actions available, and

b) There must be a valuation method by which parties can compare the value of potential exchanges.

For example, how much will a process designer be willing to increase the costs of production to allow for an additional design feature? The product and process designers must have the discretion to make this change, and they must have some way of determining whether the exchange is worthwhile for one or both of them. The only valuation method that will guarantee that decentralized decisions are made in accordance with the global objectives is for each agent to share the global objective. This will guarantee that goals among peers are consistent with each other and consistent with the global objective. If either goal consistency among peers or congruence with global objectives is missing, then bargaining agents may never agree on anything, or may agree on the wrong design solution, causing a failure of the market mechanism to coordinate the functions properly. Goal consistency and congruence were identified by Vroom and Yetton (1973) and Williamson (1975) as essential for decentralized decisions to be coordinated by market mechanisms.

In competitive markets with many competing players the global value of any individual action is efficiently communicated in the market price. In a design project the price of an action is its effect on the global objective. Determining this price is nontrivial, typically involving experimentation and additional overhead expense. Also, due to interdependencies across groups, prices will change dynamically in time as each group evolves its actions and design policies. The entire system is one of each group seeking constant improvements by dynamically adjusting to actions of other groups. As any one group makes a change, the consequences ripple through the organization as each group responds to the new reality. In summary, the hallmarks of market-based integration are decentralization (deploying discretion to groups), a valuation system for each group to focus its discretion (which must be as close to the global objective as possible for the system to be effective), and continual mutual adjustment as the entire system evolves toward an optimal or satisfying design.

Constraint-based integration: A constraint-based method relies on rules or other restrictions on behavior to achieve the desired integration. Each action by each group is constrained to follow a preset pattern, so that coordination is guaranteed by organizational design. The archetype is a perfect bureaucracy, which has preset contingency plans that prescribe appropriate actions for any circumstance, and ideally works automatically like a programmed machine (*cf* Merton 1940). The hallmark of a constraint-based integration plan is lack of discretion, wherein the actions of each group are constrained to follow preset plans. The constraints are chosen so that if all groups stay within their constraints there will be no inter-group conflict. This reduces the need for bartering and negotiations between groups, lowering the overhead expense of project coordination.

The distinction between market-based integration and constraint based integration can be illustrated as follows. Consider a web connecting each task to each other task in a project. The market model visualizes the actions of any one group reverberating throughout the web in the form of price adjustments, and each other group dynamically adjusts its activities according to prevailing prices. The process is one of constant mutual adjustment. The constraint-based model partitions the web with rigid boundaries, so that the actions of any one group cannot reverberate outside the boundaries of its cell. This maintains the independence of groups as long as each acts within their constraints. The major obstacle, of course, is in prescribing the appropriate constraint set for each group.

For example, modular product design uses constraints at the interfaces between modules to reduce negotiation and conflict (*cf* Ulrich and Tung 1991). As long as each module is designed subject to established interface constraints, the modules will fit together and function appropriately. Module design teams need not negotiate with one another, and can exercise discretion in designing their specific module as long as they satisfy the constraints at the interfaces.

Setting constraints may appear easier than setting prices (or equivalently, providing a valuation technique to assess tradeoffs), but constraints imply that some important tradeoffs have already been decided. By denying discretion a priori, constraints limit the set of design solutions that groups are allowed to consider, and any remaining discretion can at best be used to optimize the design locally around the product architecture imposed by the constraints. Since the best product architecture is seldom known at the beginning of a project, a priori constraints may overly restrict a design effort.

Values-based integration: Values-based integration relies on social norms and culture to promote coherent actions. The idea is to induce group cohesion as much as possible by enculturating all members into a shared sense of values. While its more qualitative and less systematic in nature than either market or constraint-based techniques, some authors credit the recent success of Japanese firms to this form of socialization. Japanese manufacturers are perceived as being able to cultivate and maintain a more cohesive work culture than their American counterparts, and this greatly lowers the overhead expenses of coordination and monitoring performance. Shared values and culture have been identified with effective organizations in both the popular and scholarly management literatures (*cf* Peters and Waterman 1982; Collins and Lazier 1992; Ouchi 1980, 1981; Lawrence 1985).

The values approach is similar to the market mechanism in that it envisions all groups moving dynamically and in concert, but with no need for explicit price signals to coordinate their activities. Mutual adjustment results because each individual shares a common set of values that point to the appropriate actions in any circumstance. Values enthusiasts distinguish their brand of integration from alternatives primarily on the degree of necessary quantification. The hallmark of values-based integration is trust; trust that each individual will selflessly pursue the global community objectives.

A values-based system incurs low overhead costs once the values are in place (enculturation is complete). The problem, of course, is to define a set of values that all members of the community will subscribe to. This may be impossible, or to the extent that it is possible the prescriptions (e.g. serve your customer) may be so generic as to provide little real guidance when hard tradeoffs are necessary.

All real systems are hybrids of these three types of integration, the differences being in the relative weights placed on each. For important work habits and design features that are difficult to quantify, qualitative values can be used along with negotiations and rules to supply integrating energy.

1.2. Conflict resolution. As it may be impossible to get a group to agree completely on anything, complete goal consistency and congruency may be a noble but unrealizable goal. In a market system this limitation can result in lengthy, inconclusive negotiations. In a constraint-based system, what happens if the constraints are infeasible? For example, a design team may adopt the global agenda of designing a product with superior features at very little cost in a short amount of time. While clearly desirable, this may be impossible.

These problems have been considered by organizational theorists and transactions economists, who observe that one solution is to formally grant to a specific person or group the authority to resolve these issues when they arise. The implicit assumptions are that the authority is willing and able to reach a decision. The positioning of authority can take several forms (*cf* Galbraith 1973).

Hierarchies: Hierarchies are implicit in a conventional organizational chart. Regardless of how the subordinate teams are coordinated, when an exception occurs that they are not equipped to handle the issue is referred up the hierarchy to the manager. There are many variations on this theme. Vroom and Yetton (1973) propose a continuum from decentralized through consultative to centralized systems. The first is characterized by negotiation among the work teams, the second involves discussion among the work teams with managers facilitating the process, and the third involves dictates from the management team. Which of these is most appropriate depends on the extent of the goal inconsistencies and incongruence among the work teams (such differences will favor centralization of decisions) and the extent of the information asymmetries between subordinate teams and managers (decisions should be placed where knowledge is greatest). Completely decentralized systems require goal congruence to be effective. Completely centralized systems are equivalent to the management team removing discretion from the work teams and replacing it with constraints on their actions.

Sequential authority: Rather than creating an additional administrative layer, a sequential system specifies an authoritative ordering among work teams. Higher authority teams can impose constraints on lower authority teams, so that no interteam conflict lasts for long. The phased approach to product design has sequential characteristics, in that ideation, product engineering, process engineering, and promotions and distribution all occur in chronological sequence, and upstream processes can impose constraints on downstream processes. If each team has local rather than global sympathies, this can result in the over-the-wall syndrome. See Eppinger et. al. 1990 for a discussion of sequential procedures that minimize the negative consequences of constraint propagation with localized agendas.

No real system will be a single pure type. Markets, constraints, and values may all play a role in pursuing integration, and the authority structure may combine aspects of hierarchical and sequential control. While pure market systems with no hierarchies are commonly held up as ideals, few real systems approach the pure market form. Why this is so will be investigated in the next section. We take as a point of departure the common managerial control tactic of hierarchically dividing labor and deploying objectives, and we then look for reasons why a manager or management team may not choose to deploy complete discretion and the global objective to subordinates.

2. Managing the design process

The most common managerial control technique used in business today is a hierarchical propagation of divided labor and deployed objectives. Management by objective (MBO, *cf* Horngren and Foster 1987) is an example of this process. A manager, holding some global agenda, subdivides the labor among his or her subordinates. The manager then meets with each subordinate and outlines personal goals for them for the coming review period. The subordinates take their personal goals and further subdivide and deploy them to their subordinates, etc. Quality Function Deployment (*cf* Sullivan 1986, Hauser and Clausing 1988) is an example of this sort of process. Here we will consider a two level hicrarchy, composed of a single manager or management team, who has conflict resolving authority and is responsible for managing multiple subordinates. The management team has an explicit agenda, which we will call the global agenda, and we will assume that the management team behaves as a unitary actor. The job of the management team is to organize the subordinates to attend to this agenda. In multiple level hierarchies, each level can be viewed as an image of this general process. A formalized description of this process involves:

- Articulating a goal or agenda in terms of primary variables.
- Disaggregation of primary goals into causal actions and initiatives.
- Evaluation of the relationship between the causal actions and the primary variables (hence the objective), the division of labor and deployment of subagendas.

For example, the global objective could be a combination of relations and constraints concerning product design features like cost, product functionality, timing, and style. These could be disaggregated into specific component requirements each of which inherits cost/functionality tradeoffs, timing and budget constraints, etc. Each of these will be disaggregated further, etc. When the number of disaggregated variables becomes too large for a single person or team to handle in a timely fashion, labor is divided and each subtask receives its own specific agenda and constraints. Subtasks are then projects unto themselves, and the process of disaggregation and subdivision can begin anew. Below we discuss each step of the process in turn.

2.1. Articulation of project goals and agenda in terms of primary variables.

An explicit articulation of goals and objectives must be present at the beginning of a design project, or else the entire management structure goes unguided. If the management team does not articulate goals clearly, then several things can occur, none of them good. The project may wander in random directions from one subordinate activity to another, so that random product configurations appear before the management team for their comment and probable disapproval. Alternatively, some politically strong group will take charge and design the product as they want, probably with more attention to parochial sympathies than customer needs (*cf* Anthony and McKay 1992). The clear articulation of goals is one thing that virtually all organizational commentary specifies as necessary for focussed action.

Formally, we will assume that the global agenda can be articulated as a function of attributes plus some constraints: $Max\ f(x)$, subject to the constraint $x \in \Omega$, where x is a vector of design and project attributes such as timing, cost, functionality, and style. A representation such as this, in terms of an objective function and constraints, should always be available. The function $f(x)$ reveals the tradeoffs one is willing to make; x is welcome to change in any feasible fashion that increases f. For example, f may specify by how much cost is allowed to increase to achieve improved functionality. The price for a local change in x will be the gradient of f with respect to x as long as no constraints are violated. The function f can be evaluated from a combination of customer and firm preferences using decision sciences or marketing techniques (*cf* Cohen 1978, Shocker and Srinivasan 1979). If no function is articulated, the objective will just be in terms of constraints on timing, cost, functionality, and style. In this case the global objective is to meet those constraints.

2.2. Disaggregation of primary variables into causal variables. At the level of the management team x is specified in terms of aggregate properties like total cost, time to product launch, and total product performance. These aggregate properties, however, are the result of many smaller and more specific tasks. These tasks, activities, etc. are called the causal variables, which we will represent by another (usually significantly larger) vector a. Negotiations over the horizontal and vertical axes in a QFD matrix are examples of brainstorming to try to disaggregate primary variables (the customer requirements in a QFD exercise) into causal variables (engineering characteristics).

The aggregate variables x are assumed to be (probably complex) functions of the causal variables a. We will represent this cause/effect relationship by the function g, so that $x = g(a)$. This function is necessary to facilitate action, since purposive action requires that the consequences of one's actions be understood. However, this relationship is not typically known a priori. One of the important features of a design project is that the function g must be discovered by active investigation.

2.3. Evaluation of the relationship between causal and primary variables. Determining the relationship between x and a is typically done with mathematical or physical models, judgment, or experimentation. For example CAD, Taguchi, and prototyping techniques each involve a sequence of what if experiments that will yield some data points in the cause/effect relationship between a and x. Because time and cost are themselves constrained, typically just a few data points will be investigated, so that the complete cause/effect relationship may never be known. Still, judicious selection of which prototypes to build, etc., can yield sufficient insight to find a satisfying product given the costs of experimentation.

Complete disaggregation of even a modestly sized project will generate an explosion of causal variables. When this explosion generates a critical mass of variables, it becomes more efficient to divide the responsibility for the variables among several subordinate teams. Each such team can take charge of adjusting, evaluating, and further disaggregating the variables in their charge. In this way variables are processed in parallel, and much more rapidly than they would be if a single, boundedly rational person tried to manage them all simultaneously.

2.4. The division of labor and deployment of objectives and constraints. The division of labor is equivalent to dividing up the causal actions a into disjoint partitions, and giving authority over each partition to a separate team of individuals. Then each subordinate team is given a team objective, and the process of disaggregation, partition and deployment can begin anew within each team. The first question is who should divide the labor, the management team or the subordinates themselves?

Advantages of letting subordinates divide the labor among themselves: An overhead minimizing division of labor is one that minimizes interdependencies between subordinate teams. The true interdependencies, however, are often only apparent after extensive disaggregation of primary variables into their causal antecedents, and the evaluation of the cause/effect relationship that links them. Since this disaggregation and evaluation are typically performed by subordinates, they will be the most knowledgeable about the relevant interdependencies and are the most qualified to divide the labor intelligently.

Advantages of the management team dividing the labor: There are two basic advantages to the management team dividing the labor. First, while subordinates, in the aggregate, may be more able, the management team may be more willing to embrace the global objectives (see below) and so may be more willing to divide the labor along lines that make global sense. Second, if the correct division of labor is apparent a priori, then negotiation and experimentation time and cost can be saved if management specifies that division by fiat. The correct division of labor is one that minimizes interdependencies between subtasks. The process of the disaggregation of x into a and the evaluation of the cause/effect relationship $g(a)$ is often labor intensive, and an a priori division of labor can reduce this burden by orders of magnitude. Consider, for example, a causal vector a with nine components $a=(a_1, a_2, \ldots, a_9)$, each of which can take on five different values. To fully explore the consequences of each possible a vector will take $5^9 = 1.95$ million different experiments. If these can be divided up into three groups, $\{a_1, a_2, a_3\}$, $\{a_4, a_5, a_6\}$, $\{a_7, a_8, a_9\}$, which interact only weakly, then each group can investigate all possibilities with only $5^3 = 125$ experiments for total of 375 experiments. Since time and budget constraints will restrict the number of experiments that will actually be run, an a priori division of labor may induce a more efficient use of that exploratory effort if it minimizes intergroup dependencies.

Given any division of responsibility, what objective should each subordinate team in the divided whole get? In fact, the division of labor and the deployment of objectives are not independent activities. For example, if constraints are deployed to minimize interdependencies (e.g. modular design) then labor should be divided by module, and vice versa. Typically, functional groups with deep expertise within their functions will exist in a firm, and it may make sense to leverage that depth by dividing labor by function to some extent. However, if cross-functional interdependencies remain, as they typically do, then some means for attending to these must be considered. If goal congruence is important, and virtually all authors on this topic agree that it is, then it would appear appropriate to deploy the global objective to each subordinate team.

The cause/effect relationship g between a and x allows us to write the global objective and constraints as functions of causal variables a instead of aggregate variables x. With a slight abuse of notation we will rewrite the global objective as maximizing $f(a)$ subject to $a \in \Omega$. Deploying the global objective means that, for example, subordinate team i is told to choose the actions over which it has discretion, a_i, so as to maximize the global objective $f(a)$ subject to the contraints $a \in \Omega$. Note that each subordinate team is then asked to embrace an objective that depends on the entire action vector a, and not just on its actions a_i. This may stimulate negotiation across teams, but may also be dysfunctional, as outlined below.

Advantages of deploying global objective: If each subordinate team "receives" the global objective, then goal consistency across teams, and goal congruence with the global objective, are enhanced. This facilitates bargaining across teams, and there is a good chance for teams in concert coming to closure on a good design. In short, this facilitates decentralized, concurrent decision making and successful market mechanisms. In fact any departure from the global objective will be costly, in that any compromise of goal consistency and congruence will induce more conflict, and result in poor designs (from a global perspective) and/or more decisions being referred back to the management team for resolution.

Disadvantages of deploying global objective: The motivational consequences of deploying a global agenda, but only deploying authority over a subset of causal actions, are problematic. In the economics literature (*cf* Milgrom and Roberts 1992), the human resources literature (*cf* Walton 1985) and discussions of management control systems (Horngren and Foster), the extent to which performance can be successfully incented decreases as the correlation between local actions and the performance measurement decreases. The appropriate level of aggregate performance on which to base rewards is an often cited problem in management. For example, should a firm pay bonuses to workers based on individual, team, or firm performance? The tradeoffs are clear: higher levels of aggregation induce more goal congruence, but decrease the correlation between local actions and the reward received. For example, a single worker may see little relationship between unilateral actions that he or she may take and the total annual profits of a firm.

The ideal solution would be to deploy local incentives that do not compromise goal congruence. This is available if the objectives and constraints are separable (additive). A division of labor into n subtasks is a partition of a into n subvectors a_i, $i = 1$ *to* n. Suppose that for some such division of labor there exist n functions and constraints (f_i and Ω_i) such that

$$f(a) = \sum_{i=1}^{n} f_i(a_i),$$

and $a \in \Omega$ if and only if $a_i \in \Omega_i$ for all i. Then the global objective is

$$Max \sum_{i=1}^{n} f_i(a_i) \quad s.t.\ a_i \in \Omega_i \ \ for\ \ i-1\ to\ n$$

and we can divide the labor into n tasks, giving task i hegemony over actions a_i with the local objective to maximize $f_i(a_i)$ subject to $a_i \in \Omega_i$. Then local discretion over each a_i exercised by each subtask independently will be equivalent to the global agenda. Thus different subtasks can proceed autonomously with no interdependencies. This is seldom the case, but for specific action vectors it may be nearly true. That is, for certain design configurations *a* the design problem may be nearly separable in a neighborhood of *a*. So adding the additional constraint that the product design must be in that neighborhood facilitates separability. This is yet another perspective on modularity, where interfaces (interdependences) are stabilized and constrained, and discretion is granted to each subtask subject to the interface constraints. As noted above, this technique preordains a specific product architecture but can allow the project to go on to seek a local optimal design in the neighborhood of that architecture.

While efficiencies are available when one can exploit separable features of the global objective, typically any division of labor will leave many interdependencies among the subordinate tasks. Goal congruence, and the move toward decentralized decisions, concurrent design, and market mechanisms will in that case be enhanced by deploying the global objective to all subordinate tasks. Yet, in general, this is not done. As noted above, one possible reason is the demotivating effects of deploying a global agenda with only limited action discretion. We now expand on this, and other possible deterrents to deploying the global objective.

Organizational researchers and transactions economists agree that to facilitate a productive market system among subordinate teams, subordinates must be both willing and able to embrace the global objective and pursue it via negotiations. That is, subordinates must be willing to adopt the global agenda as their own, must have at least as much information and knowledge as managers, and must be able to leverage that knowledge in negotiations. Impediments to deploying the global objective, and hence to inducing market mechanisms, are essentially anything that reduces subordinates' willingness or ability to pursue it. Possibilities include:

1) *Extrinsic goal incongruence*: Goal incongruence can be introduced by competing agendas. If, for example, subordinate teams owe strong allegiances to their functional areas in addition to their responsibilities on the design project, then goal incongruencies will be introduced. The term extrinsic refers to goal incongruencies introduced by explicit organizational structures. It follows that these can be defused by organizational changes.

2) *Intrinsic goal incongruence*: Intrinsic goal incongruence refers to the demotivating nature of decreasing the correlation between actions and the global performance measure, for example due to information overload and/or an individualistic aversion to effort. This phenomenon depends on human psychology and is independent of any explicit organizational structure. Intrinsic goal incongruence will imply a fundamental tension between the division of labor (restricting the actions over which any one group has

discretion) and the maintenance of goal congruence across groups. Divided labor will have to be accompanied by a winnowing of the information processed by each group, and by necessity this will be an approximation to the richness of the actual interdependencies.

3) *Communication technologies*: The communication technologies available to a dispersed group of a hundred or so designers may not support active bargaining and negotiation. A management team of six to ten people may simply be a better and more complete forum for communication than a dispersed group of 100 people. The power of intimate interaction is attested to by advocates of collocation, the placing of interdependent workers in a single room or building. Face-to-face interaction provides a richer context for communication than interacting through electronic mail, for example. If the enhanced communication in a small management team more than compensates for any information asymmetries between management and subordinates, it may be efficient for the management group to make many of the important design decisions and constrain subordinates to behave accordingly.

4) *Negotiation skill differentials*: The speed and efficiency with which negotiated settlements can be arrived at depends on the interpersonal skills of the participants. The more people are enculturated with confrontational tactics, the less efficiently they will be able to reach agreement even if no theoretical impediment exists. If the management team has better negotiating skills than the subordinates, then there is an argument for the management team reaching design agreements and deploying the appropriate constraints to subordinates, even if in the aggregate more knowledge is resident at the subordinate level.

5) *An optimal or near-optimal design is known*: If detailed disaggregations and the cause/effect relationships between detailed tasks and primary variables are completely known, then optimal decisions can be identified with an analytical search rather than active negotiation. In this case there is no reason to spend time bargaining, and it is most efficient simply to constrain each person to behave according to the optimal program. Any deviations are costly so no deviations are allowed. This is the purely bureaucratic solution. This corresponds to the complete capture of all relevant subordinate knowledge in organizatoinal memory, which can be accessed by the management team. The more this information is shifted from subordinates to managers, the more efficient it is for the latter to prescribe actions instead of allowing discretion and bargaining.

Milgrom and Roberts note that action constraints can efficiently replace market prices in those instances where the optimal is known with some confidence. As an example, consider the difference between an automobile design team and an automobile assembly team in the amount of discretion allowed the lowest level workers. On a design team they still have discretion and problem solving responsibilities. On the assembly line virtually all discretion has been removed. In the latter, a combination of design models and accumulated experience through repetition results in a nearly optimal configuration, which is then encoded in constraints on behaviors.

As suggested above, current thinking in product design favors the development of decentralized tecniques, with their trademarks of concurrency and mutual adjustment. To realize this organizational objective, the above impediments to this effort must be defused. The following section suggests the relationship between some current product design techniques and the effort to decentralize decisions.

3. Design initiatives affecting decentralization

A range of current design initiatives can be placed into the above framework, some of which are reviewed here.

1) *Organizational and incentive issues*: Extrinsic goal incongruence can be reduced by formal organizational measures, and such a reduction is a necessary prerequisite for successful delegation of authority. Alternative organizational designs for product development range from a strong functional organization, through a matrix structure with a project leader of varying strength, to the extreme of a dedicated team (tiger team) with no external obligations (*cf* Hayes and Wheelwright 1988, Wheelwright and Clark 1992). Naturally, the fewer external obligations workers have the more extrinsic goal incongruency is reduced. Related to this is the adjustment of formal performance measures to accommodate project performance. For example, charging the time a manufacturing manager spends in design meetings to manufacturing overhead, rather than as a dedicated design expense, will make traditional accounting ratios look worse, and demotivate him or her to invest in the process (*cf* Rosenau 1990).

However, intrinsic goal incongruence is less responsive to formal organizational adjustments, and there is evidence that this source of goal incongruence is significant. Conventional wisdom is that there is an upper limit (typically around six to ten individuals) on the effective size of teams, beyond which one begins to lose commitment and team identity. This is a strong empirical indication that intrinsic goal incongruence is a standard feature of cooperative efforts. The consequences of this are unfortunate, because they place an upper limit on the number of people that can be coordinated by purely market-based interactions, wherein each member is simply given the global agenda and the team is left to its own devices. Beyond this size, localized agendas of reduced complexity are necessary and these may by necessity neglect some important interactions. Whether intrinsic goal incongruence is a permanent feature of large cooperative efforts, how it depends on individual psychology, and managerial methods for defusing it are not well understood at this time (but see point 8 below).

2) *Communication technologies*: Information systems that can broadcast global information to all local design teams, keeping all parties updated on the current status of all groups, are one subject of ongoing research and development in the design area (*cf* Watkins 1990, Cutkosky 1991, and references therein). These technologies can reduce the reliance on face-to-face contact, and in the absence of goal incongruence can generate a computerized market through which design changes are negotiated. However, as noted above, if each subtask does not share the global agenda, then these technologies must be accompanied by complementary managerial initiatives that mitigate goal incongruence. Deploying global information with the mandate to "take it all into account" is unlikely to be successful, since in the presence of goal incongruencies advanced information systems will likely just accelerate the rate of conflict and deadlock among teams.

3) *Quality Function Deployment (QFD)*: QFD provides a useful format for articulating goals, and for brainstorming over the appropriate disaggregation of the primary variables into their causal antecedents. Both of these are necessary to manage the design process. However, QFD does not suggest ways to divide the labor or deploy objectives. In the literature (Sullivan, Hauser and Clausing) the process is usually presented as one of deploying targets (that is, constraints). It should be noted that deploying targets only will have the following consequences. First, subordinate teams are not given sufficient information to allow them to make tradeoffs among the constraints. That is, suppose both performance and cost targets are deployed, but that both are not simultaneously attainable. The subordinate team does not have the information it needs to decide to compromise performance slightly to achieve the cost target, or vice versa. A likely consequence of deploying constraints only is that the constraints are, at best, merely met. There is no incentive to improve the product further.

Some authors (*cf* Ross 1988) suggest that Taguchi methods can be appended to the QFD targets to provide the necessary tradeoff criterion between product attributes. However, the additive, quadratic loss function used by Taguchi may be inconsistent with global objectives. If managers have specific tradeoffs that they are willing to make (which ideally should derive from customer preferences), they should be explicit about them instead of relying on subordinate teams to retrofit an exogenous tradeoff structure onto deployed targets. While his quadratic loss function is problematic, Taguchi has made other significant contributions to design efforts.

4) *Taguchi methods*: Taguchi's techniques (*cf* Taguchi et al 1989, Barker 1990) use statistical experimental designs to help researchers get the maximum amount of information out of a limited experimental budget. Thus, these techniques help subordinates get a better handle on the cause/effect relationship between causal and primary variables. As with all experimental techniques, Taguchi's methods will likely enhance decentralization, as described below.

5) *CAD, rapid prototyping, and other experimentation techniques*: Experimentation techniques are those that facilitate the efficient generation of what if scenarios. That is, for any specific action vector a the effect in terms of primary variables x is determined, giving the researchers a single pair of points in the $x = g(a)$ relationship. The easier it is to generate such pairs, the better the function g will be understood and optimal design actions identified. Because increased knowledge accrues to those who conduct the experiments, these parties will inherit a natural claim to influence design decisions. Since subordinates will conduct most of the experiments, these techniques will increase subordinate knowledge and enhance the prospects for decentralizing design decisions.

6) *Capturing knowledge in models*: Just as experiments generate knowledge, models capture knowledge and make it transportable. Since increased knowledge will facilitate the intelligent use of discretion in a design process, models can encourage either decentralization or centralization depending on who uses the models. If an analytical accumulation of detailed design tradeoffs is available to the management team, then an algorithmic search for a globally optimal design can replace decentralized negotiations. In this case, the management team will be able to deploy constraints that ensure optimal behaviors by subordinates. This sort of centralization has been the goal of much past and current activity in the fields of Operations Research and Management Science, and results in a bureaucratic system with the removal of discretion from subordinates. This is also the end result of knowledge accumulation in March and Simon's evolutionary model. In contrast, giving models to subordinate teams can increase their capabilities and encourage decentralization.

An interesting consequence of intrinsic goal incongruence is that static price models may be employed. In static price models the consequences of local actions are unaffected by actions of other groups. This increases the correlation between local actions and perceived consequences. Boothroyd and Dewhurst's design for assembly (DFA, see Boothroyd and Dewhurst 1983) models are good examples of static price models, because the computed time to assemble a product is predicted on the basis of idealized relationships--between part configuration, etc. and assembly time -- that do not change dynamically over the span of a design effort (even if process designers are concurrently changing the assembly equipment). Static price models inject some global sympathies into the local computations, but maintain autonomy. How successful they are depends on the accuracy of the static prices. Ulrich (1991) notes that the naive use of DFA models may minimize the number of parts, but can make part geometries complex and drive up mold cost and time to delivery. Snap fits, also encouraged by DFA models, are difficult to undo, making after-sales servicing difficult.

7) *Team building and negotiation skills*: Communication and negotiation training, and team building exercises are designed to minimize skill differentials between the management and subordinate teams, increasing the subordinates' ability to execute the global agenda and enhancing the prospects for decentralizing decisions.

8) *Values-based methods*: Values-based methods are aimed directly at the intrinsic aspects of motivation, since they attempt to cultivate a shared perspective that will allow each individual to suppress private preferences and embrace the common good (enhancing goal congruence). These methods are qualitative in nature, depend on issues of culture and emotional perception, are not well understood, and are treated with some skepticism by quantitatively oriented managers. However, authors who promote values-based integration do have some constructive suggestions on how to cultivate shared values. Careful articulation of goals, in a qualitative sense, is followed by consistency of action to support those goals (*cf* Collins and Lazier, Ouchi, Lawrence). It is now common for firms to adopt a mission statement that is intended to articulate the overall objectives of the firm, but is not so specific as to be quantifiable. Internal consistency requires that if an individual acts in accordance with the declared principles, then that person should be congratulated, or at minimum not penalized. In one sense, checklists and structured brainstorming (for example Failure Modes and Effects Analysis, *cf* Evans et al 1990; Value Analysis, *cf* Schonberger and Knod 1991; and empirical guidelines, *cf* Daetz 1987) are similar to values-based methods, because they are statements of broad global objectives, leaving the detailed assessment of tradeoffs to the common sense of individual designers. For example, stressing that a designer should seek to deliver the desired functionality at the lowest possible cost is a

qualitative statement that, although it can generate quantitative experiments, sets a tone rather than explicitly stating a tradeoff between cost and function. These are like mission statements, and to be effective they must be right for the project. If a design team uses its judgment to increase reliability at increased cost, and this decision is not applauded, then the force of qualitative instructions that stress reliability will be undercut. That is, a mission statement for a design project must be crafted with care and strongly supported or else it will be perceived as superficial.

4. Summary and conclusions

Many techniques for making the design process more efficient can be seen as promoting the decentralization of decisions and a move to market-based integration. Because in the aggregate subordinates will typically have more relevant information than management, decentralization would seem to make sense if this knowledge can be successfully leveraged. Impediments to realizing market-based methods involve an inability or unwillingness on the part of subordinates to embrace the global agenda. A variety of techniques has been proposed in the design literature for removing some of these impediments. These include organizational redesign, communications technologies, discovery (via experimentation) and documentation (via models) of interdependent effects, and team and culture building. Perceiving the management of a design process as the hierarchical propagation of divided labor and deployed objectives provides a framework for understanding where each of these pieces fits into the organizational whole.

An important consideration in how effective these techniques can be is the matter of intrinsic goal incongruence, which if not defused will imply an upper limit on the complexity of a project that can be pursued by market mechanisms, regardless of the extrinsic efforts taken to align goals. The presence of intrinsic goal incongruence creates an unavoidable tension between the division of labor and the breadth of the agenda that can be deployed to each partition in the divided whole. How limiting this is, and its dependence on idiosyncratic psychologies is still unknown. Another interesting feature of intrinsic goal incongruence is that static price models may be more motivational, although potentially less accurate, than dynamic price models. Values-based methods, although not well understood, attempt to address the intrinsic aspects of goal incongruence.

In the presence of an inability to deploy the global objective to each subordinate, the problem of defining which interactions to maintain and which to neglect, and who should make these decisions, remains a fundamental dilemma in organizations.

REFERENCES

Anthony, M. and J. McKay, "Balancing the Product Development Process," J Prod Innov Manag 9, 1992, 140-147.

Barker, T. Engineering Quality by Design. Marcel Dekker, N.Y. 1990.

Baron, J. "The Employment Relation as a Social Relation," J Japanese and Internat Econ 2, 1988, 492-525.

Beer, M., B. Spector, P. Lawrence, D. Q. Mills, and R. Walton. Managing Human Assets , Free Press, New York, 1984.

Boothroyd, G. and P. Dewhurst. Design for Assembly: A Designer's Handbook. Boothroyd and Dewhurst, Inc. Amherst, MA 1983.

Burns, T. and G. Stalker. The Management of Innovation. Tavistock Publications, London, U.K. 1961.

Cohen, J. Multiobjective Programming and Planning. Academic Press, New York, 1978.

Collins, J. and W. Lazier. Beyond Entrepreneurship. Prentice Hall, Englewood Cliffs, N.J. 1992.

Cutkosky, M. and J. Tenenbaum, "Providing Computational Support for Concurrent Engineering," Systems Automation, Research, and Applications, 1991.

Daetz, D., "The Effect of Product Design on Product Quality and Product Cost," Quality Progress. June 1987, 63-67.

Eppinger, S., D. Whitney, R. Smith, and D. Gebala, "Organizing the Tasks in Complex Design Projects," ASME 2nd International Conference on Design Theory and Methodology, Chicago, Sept. 1619, 1990.

Evans, J., D. Anderson, D. Sweeny, and T. Williams. Applied Production and Operations Management. West Publishing Co., St. Paul, 1990.

Galbraith, J. Designing Complex Organizations. Addison-Wesley, Menlo Park, CA, 1973.

Hauser, J. and D. Clausing. " The House of Quality," Harvard Business Review, May-June 1988, 63-72.

Hayes, R., S. Wheelwright and K. Clark, Dynamic Manufacturing. Free Press, New York 1988.

Horngren, C. and P. Foster. Cost Accounting, Prentice Hall, New Jersey, 1987.

Lawrence, P. "The History of Human Resource Management in American Industry," in R. Walton and P. Lawrence, eds, HRM Trends and Challenges, Harvard Business School Press, Boston, 1985.

March, J. and H. Simon. Organizations. John Wiley and Sons, New York, 1958.

Martin, J. Cultures in Organizations. In press.

Merton. R. "Bureaucratic Structure and Personality," Social Forces 18, 1940, 560-568.

Milgrom, P. and J. Roberts. Economics, Organizations, and Management. Prentice Hall, Englewood Cliffs, N.J. 1992.

Moore, J. "Research in Formalized Organizational Design, a Survey and Critique," Unpublished manuscript, Stanford University, Stanford, CA. June, 1979.

Ouchi, W. "Markets, Bureaucracies, and Clans." Administrative Science Quarterly, 25, 1980, 129-141.

Ouchi, W. Theory Z. Addison Wesley, Menlo Park, CA, 1981.

Peters, T. and R. Waterman, In Search of Excellence, Harper and Row, New York, 1982.

Rosenau, M. Faster New Product Development. Amacom, New York, 1990.

Ross, P, "The Role of Taguchi Methods and Design of Experiments in QFD," Quality Progress, June 1988, 41-47.

Schonberger, R. and E. Knod. Operations Management: Improving Customer Service. West Publishing Co, St. Paul, 1990.

Shocker, A. and V. Srinivasan, "Multiattribute Approaches for Product Evaluation and Generation: A Critical Review," J Marketing Res 16, 1979, 159-180.

Stoll, H. Design for Manufacture, in Tool and Manufacturing Handbook, Volume V: Manufacturing Management, R. Veilleux and L. W. Petro, editors. Society of Manufacturing Engineers, Dearborn 1988.

Sullivan, L. P. "Quality Function Deployment," Quality Progress, June 1986, 39-50.

Taguchi, G., E. Elsayed, and T. Hsiang. Quality Engineering in Production Systems. McGraw-Hill, N.Y. 1989.

Ulrich, K. and K. Tung. "Fundamentals of Product Modularity". Working paper 3335-91-MSA, Sloan School of Management, M.I.T., Cambridge, MA, September 1991.

Ulrich, K. "A Framework for Including the Value of Time in Design-for-Manufacturing Decision Making". Working paper 3243-9-MSA, Sloan School of Management, M.I.T., Cambridge, MA, February 1991.

Vroom, V. and P. Yetton. Leadership and Decision Making. University of Pittsburgh Press, Pittsburgh, 1973.

Watkins, M. "The Evolution of Structured Engineering Methodologies and Support Tools," Harvard Business School working paper, Harvard University, September 1991.

Walton, R. "Toward a Strategy of Eliciting Employee Commitment Based on Policies of Mutuality," in R. Walton and P. Lawrence, eds, HRM Trends and Challenges, Harvard Business School Press, Boston, 1985.

Wheelwright, S. and K. Clark. Revolutionizing Product Development, Free Press, New York, 1992.

Williamson, O. Markets and Hierarchies: Analysis and Antitrust Implications. MacMillan, N.Y. 1975.

14

MANAGEMENT OF DESIGN INFORMATION FLOWS

Charles M. Eastman[+] and Gordon V. Shirley[++]
University of California, Los Angeles

Acknowledgements: The authors acknowledge the assistance of Nirva Fereshatian in searching for materials used in this paper and for many helpful suggestions. The work presented has been supported in part by the National Science Foundation, grant number DDM-8915665.

Abstract: Design of human artifacts has evolved from a craftsman process in which designing was part of fabrication to a separate process involving hundreds of people, each with specialized knowledge. Complex information flows have evolved that are now significantly affecting the quality of design. A model for representing and managing design information flows is developed and some opportunities for new research outlined.

1. Introduction

Design is of growing importance in international trade. The international accounts of a nation are increasingly influenced by the quality of its products. The reputation of U.S. products has been strong for most of its history. In the early 1800s, the "American system" of manufacture led the world into mass production. At the 1851 World's Fair in London, the industrial products of the United States were praised for their utility, ruggedness and emphasis on the mass market[1]. For most of the twentieth century, American products were known for their high performance and durability. Yet today, they are judged in the marketplace and in a growing number of studies as having high unit production costs, long time to production, low reliability, poor style

---

[+] Graduate School of Architecture and Urban Planning and Department of Computer Science

[++] Anderson Graduate School of Management

[1] See A. Pulos, American Design Ethic: History of Industrial Design to 1940, MIT Press, 1983, pp. 110-118.

and inability to respond to mass markets, the opposite qualities of what were previously strengths[2].

Design is the specification of some product of human endeavor, in sufficient detail to allow fabrication. During design, attempts are made to determine with confidence the function, performance, aesthetic effect, cost and manufactur-ability, reliability, and other aspects of the final product. In striving for these objectives, design involves responding to an open-ended set of relations and involves complex trade-offs. When done by an individual, design is often considered an art form. Most often, however, design is now done by large organizations.

Today, design involves tremendous amounts of information, moving between people with specialized knowledge. The administration of these information flows is an important aspect of engineering management and a significant contributor to design effectiveness. Design information flows have changed radically during the last one hundred years, driven by new types of knowledge that improve product quality. As the amount of information has increased, design time has grown and the ability to effectively utilize the information has become more difficult. Current efforts by the Defense Advanced Research Projects Agency (DARPA) and the DOD Computer Aided Logistical Support (CALS) project and by many individual corporations are attempting to find ways to re-structure the information flow within design-engineering-production organizations, in order to improve product quality[3].

Information flow has emerged as an issue affecting design quality only recently. This paper briefly reviews design organizations and information flows, as they are currently conceived. The complexity of current information flows in design is illustrated in an example. In Section 2, design information flows are characterized by a model that allows its formal characterization, which is applied to the example. In Section 3, the importance of information flows in design management is discussed and some directions for future research identified.

---

[2] M. Dertouzos, R. Lester and R. Solow, Made in America: Regaining the Productive Edge, MIT Press, 1989.

[3] Winner, et al., "The Role of Concurrent Engineering in Weapon System Acquisition", Institute for Defense Analysis, Report R-338; Glover, et al., "An Expert R&M Design Synthesis: An Enabling Technology for Concurrent Engineering", in Computer-Aided Cooperative Product Development, D. Sriram, R. Logcher and S. Fukada (Eds.), Springer Verlag, N.Y., 1991, pp. 487-501.

2. The Current State of Design Information Flows

Until recently, the practices for managing information within a complex design organization were treated manually. They could be characterized as typically taking the following form. An initial design concept was developed by a lead team. This initial design identified the basic geometry of the product and the major subsystems. The subsystem interfaces were defined in enough detail that the overall performances of the product could be predicted, as well as estimates of its resource costs (time, materials and manpower). Also, the technical risk of achieving these performances would be generally identified. The concept design was recorded and reproduced in a few drawings and reports.

After management approval, the concept design was distributed to the various engineering teams who decomposed it and developed its specific parts in more detail. Each team represented either a functional specialty (aerodynamics) or a part of the product (wing design). Both functional and product organizations have their advantages[4]. As long as each team achieved the functional performances set for it and stayed within its geometry and met its performance limits, they could proceed independently of the other teams. If a team moved to a design outside of the specifications, then they had to coordinate with the concept team, with the changes possibly propagating to other teams. Some subsystems would be decomposed hierarchically several times, with teams working in a similar manner on each of the components[5]. Individual work was reviewed by the team leader before approval. The design was incrementally defined in a set of approved drawings. Most were reproduced and distributed to other teams needing the information. Each drawing had a completion date, a record of its author(s), and a change column.

Design information typically includes test models and many drawings. Drawings on thousands of sheets of paper were archived in a central location, often close to the drafting room; the other documents were retained by each department. Only later did the decision come about to centralize all design information in a single file or set of files.

One indication of the complexity of information flows in a modern manufacturing organization is indicated by the chart shown in Figure 1. It maps the information flows within a major automobile company, as of 1985, regarding a design change for an engine part such as a manifold, after the

---

[4] Woodson, TT. Introduction to Engineering Design, McGraw-Hill, N.Y. Chapter One.

[5] Bond, A.H. and R.J. Ricci, "Cooperation in Aircraft Design", in D. Sriram, R. Logcher and S. Fukada, op.cit., pp.152-182.

change has been released from engineering. In this chart, it is indicated that a engineering release drawing is distributed to 90 other locations, with the potential that changes may then accrue to many other drawings produced by the 90 other departments or operations. This level of complexity was achieved within a manual drawing distribution environment, just prior to automation.

2.1. The Advent of Computing. The growing availability of computers from the 1960s onward has allowed computer-based design tools to automate what were previously manual tasks. Layout and design can now be done with computer-aided drafting. Analysis techniques are automated and now could be run in a few moments, once the data was prepared, allowing them to be used much more extensively. Numerical control machining no longer required painstaking programming of cutter path trajectories, but could be generated from high level instructions.

Initially, task automation took place within traditional departmental structures, in a one-for-one substitution of automated operations for manual ones. The more practical aspects of fabrication design, integrating trade and fabrication practices, required human judgment and could not be automated. Also, electronic drafting did not address the continued use of previous paper drawn designs; there was no cheap means to convert these for electronic use. Some early proposals for integrated models of a product were made in the 1960s and 1970s, based on central data processing models, but these were limited by the performance of then available computers and the restricted support for man-machine interaction[6].

In the 1970s, computer access moved to terminals connected to a time-shared mainframe and in the 1980s, to local area networks connecting personal computers and engineering workstations with mainframes. The quick evolution of personal computers and workstations allowed them to become computationally more powerful than the largest computers of ten years earlier. Networks potentially allow all data to be stored only once, then accessed remotely from any computer on the network. Such single copy storing eliminates the need for physical distribution and reduces the problem of maintaining current updates. Local printers/plotters can make hardcopies, only if necessary.

The continued growth in computer power allows new tools to be developed, filling in between those tasks that are automated earlier. Complex phenomena that previously required model testing have eventually become

---

[6] Vlietstra, J. and R.F. Weilinga (eds.), Computer-Aided Design, North-Holland Press, N.Y. 1973.

tractable. Important examples are fluid flow simulations, new forms of finite element modeling and optimization of complex structures[7]. Automatic design rule checkers, using expert systems, began taking over certain quality control and evaluation tasks[8]. At the same time, digital scanning of paper documents and the evolving standards for large format scanned documents, allows even data that was originally on paper to be converted to an electronic format[9]. As a result, all design data is quickly moving to an electronic format. It has become possible to store all information for a project on one or a few magnetic disks. The full implications for this transition are only now being fully recognized.

Even the level of information flow shown in Figure 1 has become inadequate for the need. International competition was forcing additional levels of coordination, specifically:

- coordinating schematic design so that it incorporates manufacturing, assembly and testing considerations, so that manufacturing steps may be reduced;
- greater parallelism, especially between design and manufacturing, to reduce time to production.

2.2. Example Engineering Design Task. The flow complexity now being attempted in engineering is possibly best characterized by an example. The example below has been constructed by the authors, based on recent literature[10], to reflect the typical flows encountered in modern engineering design. It considers the design of an automobile door panel, only a small part of a larger product design. Previous work must have already defined the general body design, with detail definition of surface contours. Various sections, elevations and surfaces are generated from the computer model and passed on

---

[7] For aerodynamic simulation, see A. Jameson, "Computational Aerodynamics for Aircraft Design", _Science_, vol. 245, 28 July 1989, pp. 361-371; for design optimization, see Gero, J. (ed.), Design Optimization, Academic Press, Orlando, FL (1985).

[8] Sriram et al, op.cit, section on Design Methods.

[9] Eastman, C.M., "New view of CAD: Scanning and Revision of Engineering Drawings", J. Information and Image Manag. 19:8 (August,1986), pp. 14-17.

[10] Clark, K.B., and T. Fujimoto, "Lead Time in Automobile Product Development Explaining the Japanese Advantage", J. Engr and Tech. Manag. 6:1, (September, 1989), pp.25-58; Clark, J.B. and T. Fujimoto, "Overlapping Problem Solving in Product Development", in K. Ferdows (ed.), Managing International Manufacturing, 1989, Amsterdam: Elsiver Press.

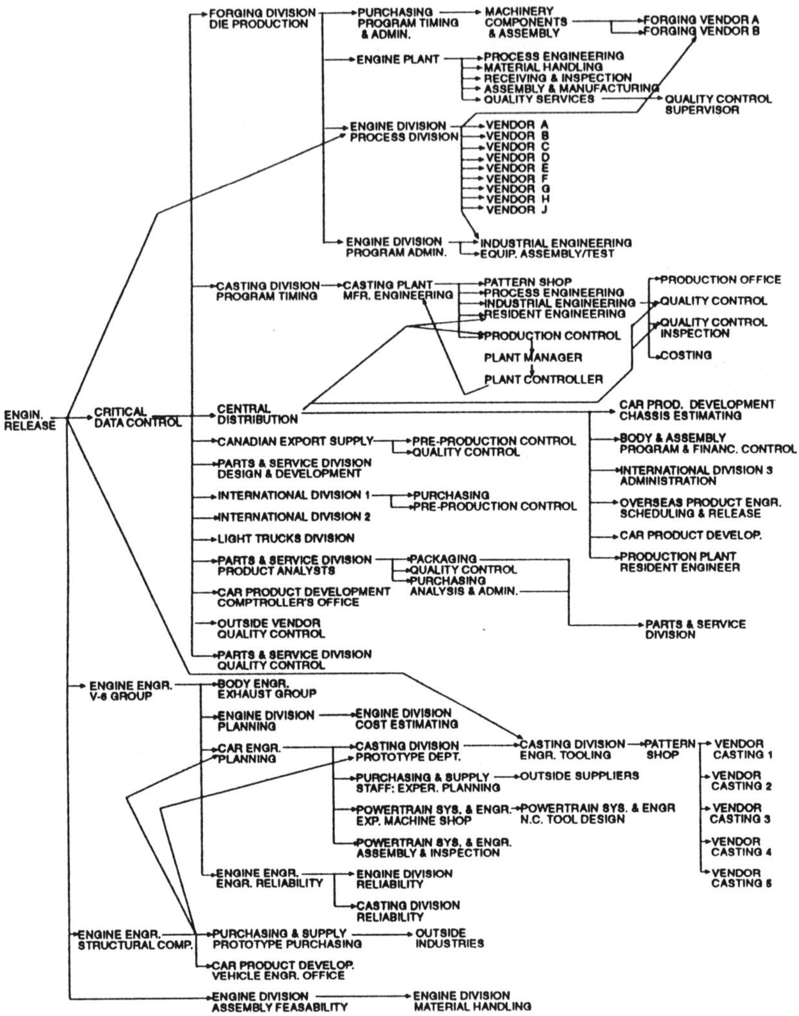

Figure 1: The information flow after engineering release of a drawing for an engine part, circa 1985.

to the detail design departments in engineering. We consider the door panel's forming and structural development, which includes an interior longitudinal beam for safety. Doors typically involve a window mechanism plus other devices: lock, venting ducts and radio speakers, but these are not considered here.

Development is considered to take place in three different departments: structures, design and process planning. Process planning develops the manufacturing schedule. The tasks and states of the design undertaken by each department and also the flows between departments are shown in Figure 2. The initial drawings and CAD database are delivered to design, which develops the initial configuration of the overall door, defining zone allocations for interior items and integrating the outside design with the major interior panel elements. After the initial configuration is developed, the various components are detailed. Here, we track three elements: the outside and interior panels and the structural beam.

Each of these elements are defined adequately for two types of analysis. The first is structural deformation under various loads. The resistance of the beam to lateral crash loads is also analyzed. The other type of analysis is metal forming, involving the stamping of the parts using dies, anticipating metal springback and consideration of different gauges of metal. These analyses are made by the structural group, using finite element models. After this work, process planning determines how the elements must be fabricated. Last, schedules for each element and the assembly of the door are defined.

Design development goes through many steps, involving many datasets or drawings. During this segment of the process, twenty-one different representations are used, listed below:

<u>DOOR PANEL DOCUMENTS</u>

E1. overall layout dataset or drawings
E2. overall door panel dataset or drawings
E3. outer panel dataset
E4. inner panel dataset
E5. outer panel flexure FE dataset
E6. outer panel metal bending finite element dataset
E7. inner panel metal bending finite element dataset
E8. safety beam flexure FE dataset
E9. outer panel die design dataset
E10. inner panel die design dataset
E11. overall door preliminary process plan

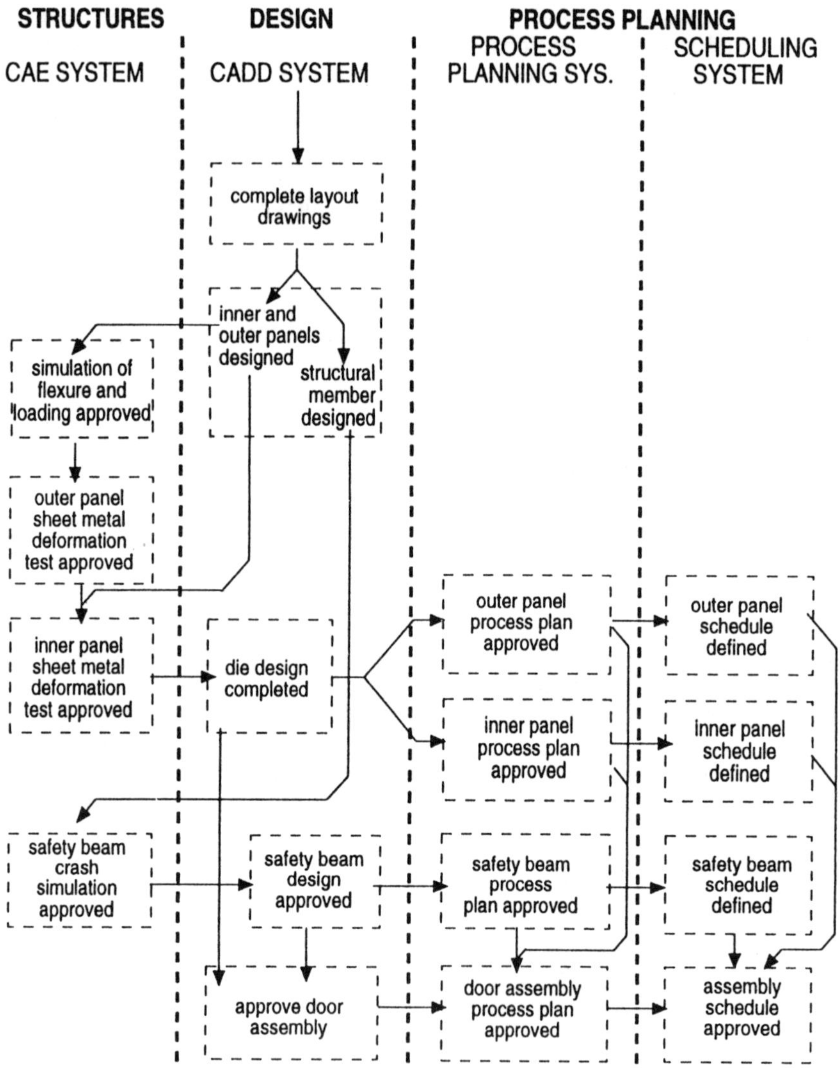

Figure 2: Task sequence for the example engineering problem.

E12. safety beam process plan
E13. outer panel process plan
E14. inner panel process plan
E15. overall door preliminary schedule
E16. outer panel fabrication schedule
E17. inner panel fabrication schedule
E18. safety beam fabrication schedule
E19. overall door assembly model
E20. overall door assembly process plan
E21. overall door assembly schedule

Several weaknesses exist in this process, however. Suppose that deformation analysis of one of the elements rejects the design. Not only must the door panel be re-designed, but the process plan work must also be re-done. To better deal with these interacting conditions, the structural work should be integrated with the component design. Also, the process shown in Figure 2 is almost serial, with only a few parallel activities. Design proceeds to structural analysis, followed by process planning.

There are several areas where greater parallelism is possible. The metal bending analysis of the outer panel and die making could be undertaken in parallel with the structural flexing analysis, based on preliminary drawings, with iterations leading to joint solution of both requirements. The safety beam structural analysis can be done after the preliminary design is defined and prior to approval of the final drawings. Several other examples of parallelism are possible. However, the planning and coordination for such levels of parallelism are very challenging.

The challenge today is to manage the information flows defined above, plus several hundred other parallel flow patterns for the different components, in a highly efficient manner. Also needed are means to generate alternative flow structures and to evaluate their effect on resources and time. The need has moved beyond fixed flows with minor iterations.

3. Engineering Information Systems

In the early 1980s, electronic file management systems were developed for storing the various electronic documents or files used in engineering and design. Called *Engineering Information Systems* (EIS), these systems allow design data that is represented electronically to be centrally stored, managed and indexed for access.

The original model for EISs was software engineering, which had similar needs[11]. The management techniques involved are called *configuration management*. In EISs, one or more files hold a dataset, such as for representing a drawing, a set of analysis input data, a 3-D surface model, analysis results, text or instructions to drive some machine. The configuration management tools within the EIS stores engineering data on a central library server (or on a network controlled by a central library system), and provides services that correspond to those required for design data management:

- *document distribution,* using networks and distributed hardcopy devices;
- *version control*, with access to the most recent version of each dataset, with a trail to previous versions;
- *access control*, with authorized data check-in and check-out of datasets;
- *recording changes* and those responsible for them, and incorporation of *an approval mechanism* before updating;
- *information management*, including approval procedures.

Approval usually needs an explicit sign-off, leading to implementation of an "electronic signature". After a user's changes are approved, the new document may be automatically distributed to those using it. An EIS may provide other services such as query access to the datasets. Some EISs provide support for managing the implications of changes, where one change propagates to several others, while others provide management reporting capabilities. Systems for such use allow hundreds of thousands of datasets and/or documents to be directly accessed, using "jukebox" optical disk systems with up to a terabyte of data storage[12].

EISs cross departmental boundaries, which pose organizational complexities in their implementation. They require modifications to the way automation and related procedures are handled in multiple departments. What had been informal procedures that could be worked out locally within a small group, are now found to have larger ramifications that are justified primarily by benefits at the division or corporate level.

The development of EIS technology emerged at the same time as the impact of foreign engineering and product design impacted sales by U.S. industries. Part of the motivation for EIS, then, has been the need to shorten development cycles, reduce iterations and speed the production of new

---

[11] See W. Babich, Software Configuration Management, Addision Wesley, Reading, MA, 1986.

[12] See Special Issue on Engineering Document Management, INFORM, 2:9, October, 1988.

products. Where manufacturing earlier was told what was to be manufactured by styling and engineering, it is now recognized that manufacturing must influence the initial design, so as to reduce fabrication, finishing, packaging and other downstream problems and costs. Testability, maintainability and assemblability are among the "abilities" that are recognized as important inputs to conceptual design.

3.1. A Model of Design Information Flow. In the mid-1980s, several formulations of the configuration management process emerged[13], allowing definition of the characteristic information flows within any design organization. Given these flows, policies regarding subsets of the flows between processes could be characterized within a particular organization. In some cases, the dependencies between design operations could also be characterized.

In the following, a general structure of design information flows is presented[14]. In order to deal rigorously with design information management, precise definitions are needed. We rely on the following[15]:

a. A document is defined as an *entity*, where e_m is document m in the set E. During design, information is added to, modified and possibly deleted from a document.
b. At various points in time, the information in the document satisfies conditions that are important to other design tasks. We consider these conditions specifiable in the form of tests and characterize them as *constraints*, denoted as c_j, for the jth constraint in the set C. A constraint is an expression that, when evaluated, has the value TRUE or FALSE, where TRUE means that the associated condition has been satisfied.
c. A document is said to transit through a series of *design states*, where a state is the different possible combinations of truth values for the document's constraints. The design state of a document, at some point in time, is denoted by the set of truth values of its constraints, e.g., the doublet $<e_m, C_m>$, where:

   e_m is the document identifier,

---

13 Neumann, T., "On representing the design information in a common database," *ACM SIGMOD Proc.* Engineering Design Applications, pp.81-87, and Katz, R., Information Management in Design, Springer Verlag, N.Y., 1985.

14 Eastman, C. and A. Kutay, "Transaction Management of Design Data", in Sriram, D., R. Logcher and T. Fukada (eds.), op.cit.

15 In the following, we use sets and predicate logic notation.

C_m is a set of current values of constraint for document m.
The state of the EIS is denoted by $U\langle e_n, C_n\rangle^t$ for all n at time t.

d. Design documents may be accessed in either of two modes: *read mode* allows the user to access the document to view and possibly extract data from it, but not change it; *write mode* allows the user to modify the document.

e. A design *transaction*[16] a_i for the ith transaction, involves accessing one or a set of documents, at least one of which is in write mode. The write document may be modified manually. Alternatively the modification may be by an application program. Other documents are accessed in read-only mode. These are assumed to be used to guide the modification or as another reference during the modification.

f. A design transaction usually requires that some information already exists and that certain conditions be satisfied for that information. For example, the application of a finite element model for structural loading requires prior definition of the geometry of the part and the external loads. The effect of a design operation is the satisfaction of certain other conditions, for example, identification of the maximum stress in the shape. Together, these conditions specify that a design transaction is of the form:

$$a_i(E^R_i, E^W_i, C^B_i, C^A_i) \tag{1.1}$$

where transaction a_i consists of a set of pre-condition constraints C^B that must be TRUE over the sets of documents E^W_i and E^R_i, a set of post-condition constraints C^A_i (that are changed to TRUE or FALSE), which apply over a set of *documents* E^W_i. Every design transaction can be characterized in this general manner, including various CAD applications, analysis packages and report generators.

g. Every document e has a unique *version identifier* v that can be ordered vis-a-vis all other version identifiers for all documents. Most often such a version identifier is a timestamp. Every document which is written, that is a member of the set E^W_i, has its current version updated at time i. From each version of a document it must be possible to reference the previous version of the document, if one exists. The versions of the documents providing the necessary initial conditions for the transaction, e.g., the documents E^R_i and

---

[16] Here, we use the term *transaction* in the sense used in databases to designate a set of actions that satisfy a set of integrity conditions (here, some constraints). This view is spelled out more fully in Kutay, A. and C. Eastman "Transaction Management in Engineering Databases", ACM SIGMOD Proceedings, Engineering Design Applications, 1982, pp.73-80.

E^W_i are also accessible. One possible implementation to versioning, then, might be that each document have an associated structure:

$$(e_{v+}, e_{v-}, s, E^D_i)_i \qquad (1.2)$$

where

$$E^D_i = E^R_i \cup E^W_i$$

where e_{v+} is the document written, e_{v-} is the previous version of the document, s is the version identifier and E^D is the set of document versions that e_{v+} depends upon. It is assumed that all documents in the sets E^R and E^W are pre-conditions for all documents in the set E^W. This corresponds to a backward reference, from the current document to the most recent previous one.

h. Generally, there is only one successor version of a document. An exception occurs when two or more documents are offered as alternatives to the previous one and the choice between them is suspended. Each document alternative may be used as a basis for later transactions. However, the design conditions are duplicated, leading to redundant work. At any time, one alternative may be selected over the other(s) and work on the other(s) terminated. This action invalidates all document generation or revisions that depend upon the terminated alternative.

Given the above definitions, any design action is feasible that satisfies its necessary pre-conditions C^B. A very large set of design transactions are possible at any one time. In practice, an organization is expected to impose a more restrictive management policy regarding the sequence of development.

It is not expected that most constraints will be defined procedurally. Rather it is assumed that each constraint is an assertion that some condition is TRUE (or FALSE) and that the assertion need only be evaluated by name for the document. For external applications, front-end pre-processors evaluate the conditions required by the application. If they are satisfied, then the application can execute. The application may terminate without completing its transaction, resulting in no change. If it does terminate, then a post-processor writes back to the file manager both the updated file and also denotes the new conditions satisfied by it.

Condition (g.) can guarantee that queries will only access the most recent version of a document, unless specifically requested otherwise. This model allows multiple documents to be merged or alternatively one document to be split into multiple documents. Approved alternatives must be managed in such a way that only one alternative remains at the completion of the design.

Separate from the transaction model defined above, an EIS also incorporates an approval structure. The approval structure, in its conventional use, defines the process by which documents may be created, extracted, modified and inserted into the EIS. Normally, each document used in a transaction goes through a sequence of states, which reflect an approval policy defined by the organization. For example, one policy defining possible statuses is shown in Figure 3. In this policy, two forms of access are possible: *checkout to read* and *checkout to write*. *Checkout to write* corresponds to a modification operation. All modifications must be approved before they may be accessed by others. More elaborate policies may be developed.

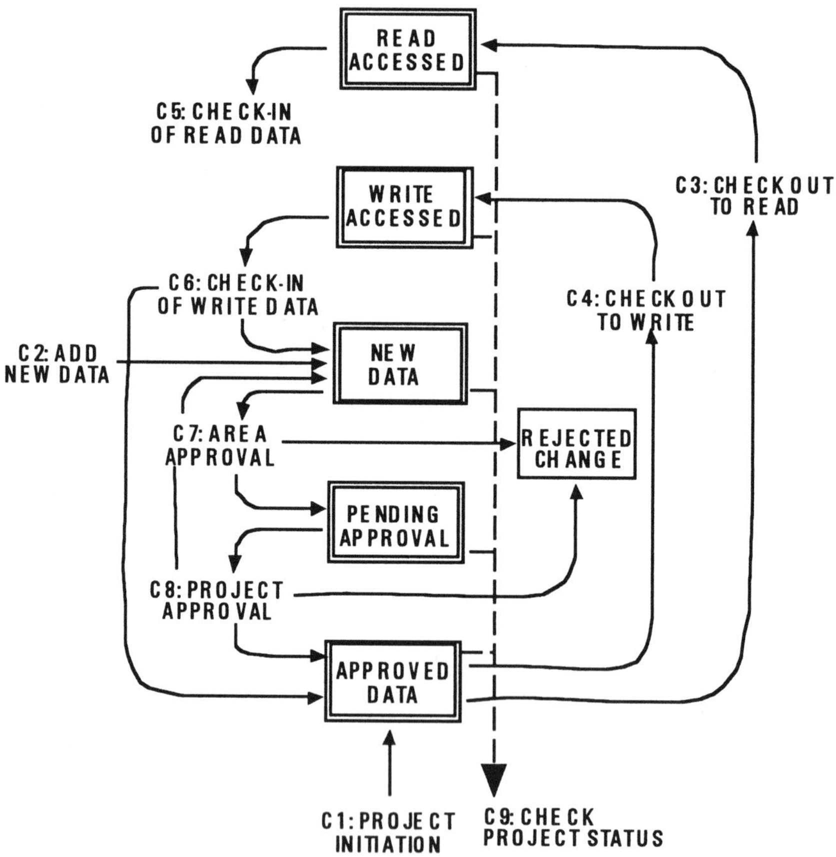

Figure 3: One possible approval policy for engineering documents.

The approval state conditions defined in Figure 3 require comment. Multiple copies of documents may be checked out to read. It is usual policy, however, that only one copy of a document may be checked out to write. Thus while a document is in the status *check out to read*, it may be checked out by others. If the document is checked out to write, then it transits to that status and no more *check out to write* transactions can be initiated. *Check out to read* is still allowed when a document is in *pending approval* state; the document accessed, however, is the previously approved one. When a new version is approved, all current readers of the document are notified. Most of these rules are policies, in the sense they are convenient; they are not logically necessary and other policies may be developed[17].

When a new version of a document is approved, it consists of the following information:

$$a_i = (E^R_i, E^W_i, C^B_i, C^A_i)$$

where

$e_{n,p,q} \in E^W_i$ and denotes version p of document n in status q,

$$e_{n,p,q} \rightarrow e_{n,p+1,q+1} \qquad (1.3)$$

Upon updating, the version is incremented as well as the project's design state. Design changes in one document may lead to changes in other documents. For example, a change in an airplane engine may require changes to the structure of the engine, its control interfaces, cowling, etc. These dependencies are naturally captured in the pre-conditions of any transaction. Thus the engine structure, control interfaces and cowling all use information from the engine documents in their design. Each would refer to the engine in its E^R and to necessary design states in C^B. Later, when the engine is changed, its version revision should automatically include flagging all the documents that referred to the modified documents in their generation. The flagged documents are to be checked for possible revision, based on the state change. This can be done as follows. All documents e_x must be flagged as a result of transaction a_i:

$$a_i = (E^R_i, E^W_i, C^B_i, C^A_i)$$

$$\forall e_x \mid (e_x \in E^W_i) \Rightarrow e_x \in E_y$$

$$\forall a_j \, \forall e_y \, \forall e_p \mid (a_j \neq a_l) \wedge (e_y \in E_l^R) \wedge (e_p \in E_j^W) \Rightarrow c_p \subset E^D_k \qquad (1.4)$$

Some of the flagged documents will be changed as a result of the initial change, resulting in a possible recursive chain of propagations. In more

---

[17] See Kung, H.T. and J. Robinson, "On optimistic concurrency control" , <u>ACM Trans. of Database Systems</u>, 6:2, pp. 213-226.

structured situations, it may be possible to automatically extract the information needed for a design transaction, for example an analysis. In such cases where the transaction is deterministic, the revision update also can be automated.

An elaboration needed in a fully functional system is to add to each document its current approval status, which is included as q above. After checking, these documents are updated in the form in eq. 1.3. If a document is changed, then they result in change propagations per eq. 1.4.

This model of engineering transactions is based upon and models design as a set of transactions that is partially ordered and which can be conceptualized as a very large precedence network. Each node is a state of the design and each edge leaving a node are those actions allowed while the design is in that state. Because most transactions can be done multiple times and in multiple alternative sequences, the precedence graph of possible actions at each state is huge and it is impractical to try to enumerate them. It is quite possible for a computer to determine what actions are currently feasible. It can also determine what documents must be reviewed if a transaction is iterated.

Given this formalized structure, it is possible to define a number of queries to an operational EIS that would be of interest. For example:

- query identifying transactions that are allowed at this time:

$$\forall a_i, \forall c_i \mid (a_i \in A(E^R{}_i, E^W{}_i, C^B{}_i, C^A{}_i), c_i \in C^B)) \wedge c_i = \text{TRUE})) \quad (1.5)$$

- query of a specific document for those constraints that it does not currently satisfy:

$$\forall s, \exists e_x \mid (U < e_x, s > \wedge\ s = \text{FALSE}) \quad (1.6)$$

- query of documents that are currently checked out to read:

$$\forall e_x \mid e_x = e_{x,p,q} \wedge (p=\text{"current"}) \wedge (q = \text{"check out to read"}) \quad (1.7)$$

Other queries can be easily formulated. These examples are not to suggest that sets and logic are an appropriate query language, but rather that the formalisms exist within this framework to develop such queries.

This formalization defines the logical structure of an EIS. It serves as a specification for an EIS, implemented on top of a database management system. This formalization has been developed for managing heterogeneous design data on a large central file manager. The granularity of management is a document (or dataset), as it is assumed that the format and logic of data inside of an application or CAD dataset is not available to an outside management scheme. This granularity matches well with the current needs in computer aided design and manufacturing.

3.2. Formalization of the Example. The automobile door panel example, developed in section 2.4 can be mapped into the EIS model outlined above. Using the formulation, we can show how the EIS model would manage the high degrees of parallelism desired.

Within the twenty-one documents presented earlier, one can identify an expanded set of thirty-four states of the design and the information flows between them, which can be used to achieve higher degrees of parallelism. The states are listed below in word form:

C1. detailed layout drawings defined (on E1)
C2. detailed layout drawings approved (on E1)
C3. door design concept defined (on E2)
C4. sheet metal components defined (on E2)
C5. safety beam defined (on E2)
C6. outer panel concept defined (on E3)
C7. inner panel concept defined (on E4)
C8. simulation of structural flexure of outer panel okayed (on E5)
C9. metal bending deformation model of outer panel okayed (on E6)
C10. metal bending deformation model of inner panel okayed (on E7)
C11. structural flexure of safety beam okayed (on E8)
C12. die design for outer panel defined (on E9)
C13. die design for outer panel approved (on E9)
C14. die design for inner panel defined (on E10)
C15. die design for inner panel approved (on E10)
C16. safety beam design approved (on E2)
C17. door panel design approved (on E2)
C18. overall door preliminary process plan defined (on E11)
C19. safety beam process plan defined (on E12)
C20. safety beam process plan approved (on E12)
C21. outer panel process plan defined (on E13)
C22. outer panel process plan approved (on E13)
C23. inner panel process plan defined (on E14)
C24. inner panel process plan approved (on E14)
C25. overall door preliminary schedule defined (on E15)
C26. outer panel fabrication schedule defined (on E16)
C27. inner panel fabrication schedule defined (on E17)
C28. safety beam fabrication schedule defined (on E18)
C29. door assembly detailed drawing developed (on E19)
C30. door assembly detailed drawing approved (on E19)
C31. assembly process defined for door (on E20)
C32. assembly process approved for door (on E20)

C33. door assembly schedule defined (on E21)
C34. assembly schedule approved (on E21)

Between these states are a set of design transactions that allow progress on completing the design. They are presented in the format shown in eq. (1.1):

<u>Design Transactions</u>

$a_1(\{0\}^R, \{E_1\}^W, \{0\}^B, \{C_1\}^A)$
$a_2(\{E_1\}^R, \{E_1\}^W, \{C_1\}^B, \{C_2\}^A)$
$a_3(\{E_1\}^R, \{E_2\}^W, \{C_2\}^B, \{C_3\}^A)$
$a_4(\{E_2\}^R, \{E_2\}^W, \{C_3\}^B, \{C_4\}^A)$
$a_5(\{E_2\}^R, \{E_2\}^W, \{C_3\}^B, \{C_5\}^A)$
$a_6(\{E_2\}^R, \{E_3\}^W, \{C_4\}^B, \{C_6\}^A)$
$a_7(\{E_2\}^R, \{E_4\}^W, \{C_4\}^B, \{C_7\}^A)$
$a_8(\{E_3\}^R, \{E_5\}^W, \{C_6\}^B, \{C_8\}^A)$
$a_9(\{E_3\}^R, \{E_9\}^W, \{C_6\}^B, \{C_{12}\}^A)$
$a_{10}(\{E_4\}^R, \{E_{10}\}^W, \{C_7\}^B, \{C_{14}\}^A)$
$a_{11}(\{E_9\}^R, \{E_6\}^W, \{C_{12}\}^B, \{C_9\}^A)$
$a_{12}(\{E_5, E_6, E_9\}^R, \{E_9\}^W, \{C_8, C_9, C_{12}\}^B, \{C_{13}\}^A)$
$a_{13}(\{E_{10}\}^R, \{E_7\}^W, \{C_{14}\}^B, \{C_{10}\}^A)$
$a_{14}(\{E_7, E_{10}\}^R, \{E_{10}\}^W, \{C_{10}, C_{14}\}^B, \{C_{15}\}^A)$
$a_{15}(\{E_2\}^R, \{E_8\}^W, \{C_5\}^B, \{C_{11}\}^A)$
$a_{16}(\{E_2, E_8\}^R, \{E_2\}^W, \{C_{11}, C_5\}^B, \{C_{16}\}^A)$
$a_{17}(\{E_2, E_9, E_{10}\}^R, \{E_2\}^W, \{C_{13}, C_{15}, C_{16}\}^B, \{C_{17}\}^A)$
$a_{18}(\{E_2\}^R, \{E_{11}\}^W, \{C_5\}^B, \{C_{18}\}^A)$
$a_{19}(\{E_2\}^R, \{E_{12}\}^W, \{C_5\}^B, \{C_{19}\}^A)$
$a_{20}(\{E_9\}^R, \{E_{13}\}^W, \{C_{12}\}^B, \{C_{21}\}^A)$
$a_{21}(\{E_{10}\}^R, \{E_{14}\}^W, \{C_{14}\}^B, \{C_{23}\}^A)$
$a_{22}(\{E_9, E_{13}\}^R, \{E_{13}\}^W, \{C_{13}, C_{21}\}^B, \{C_{22}\}^A)$
$a_{23}(\{E_{10}, E_{14}\}^R, \{E_{14}\}^W, \{C_{15}, C_{21}\}^B, \{C_{24}\}^A)$
$a_{24}(\{E_2, E_{12}\}^R, \{E_{12}\}^W, \{C_{16}, C_{19}\}^B, \{C_{20}\}^A)$
$a_{25}(\{E_{11}\}^R, \{E_{15}\}^W, \{C_{18}\}^B, \{C_{25}\}^A)$
$a_{26}(\{E_{13}\}^R, \{E_{16}\}^W, \{C_{22}\}^B, \{C_{26}\}^A)$
$a_{27}(\{E_{14}\}^R, \{E_{17}\}^W, \{C_{24}\}^B, \{C_{27}\}^A)$
$a_{28}(\{E_{12}\}^R, \{E_{18}\}^W, \{C_{20}\}^B, \{C_{28}\}^A)$
$a_{29}(\{E_2\}^R, \{E_{19}\}^W, \{C_{17}\}^B, \{C_{29}\}^A)$
$a_{30}(\{E_{19}\}^R, \{E_{19}\}^W, \{C_{29}\}^B, \{C_{30}\}^A)$
$a_{31}(\{E_{19}\}^R, \{E_{20}\}^W, \{C_{29}\}^B, \{C_{31}\}^A)$
$a_{32}(\{E_{19}, E_{20}\}^R, \{E_{20}\}^W, \{C_{30}, C_{31}\}^B, \{C_{32}\}^A)$
$a_{33}(\{E_{20}\}^R, \{E_{21}\}^W, \{C_{31}\}^B, \{C_{33}\}^A)$
$a_{34}(\{E_{15}, E_{16}, E_{17}, E_{18}, E_{20}, E_{21}\}^R, \{E_{21}\}^W, \{C_{25}, C_{26}, C_{27}, C_{28}, C_{32}, C_{33}\}^B, \{C_{34}\}^A)$

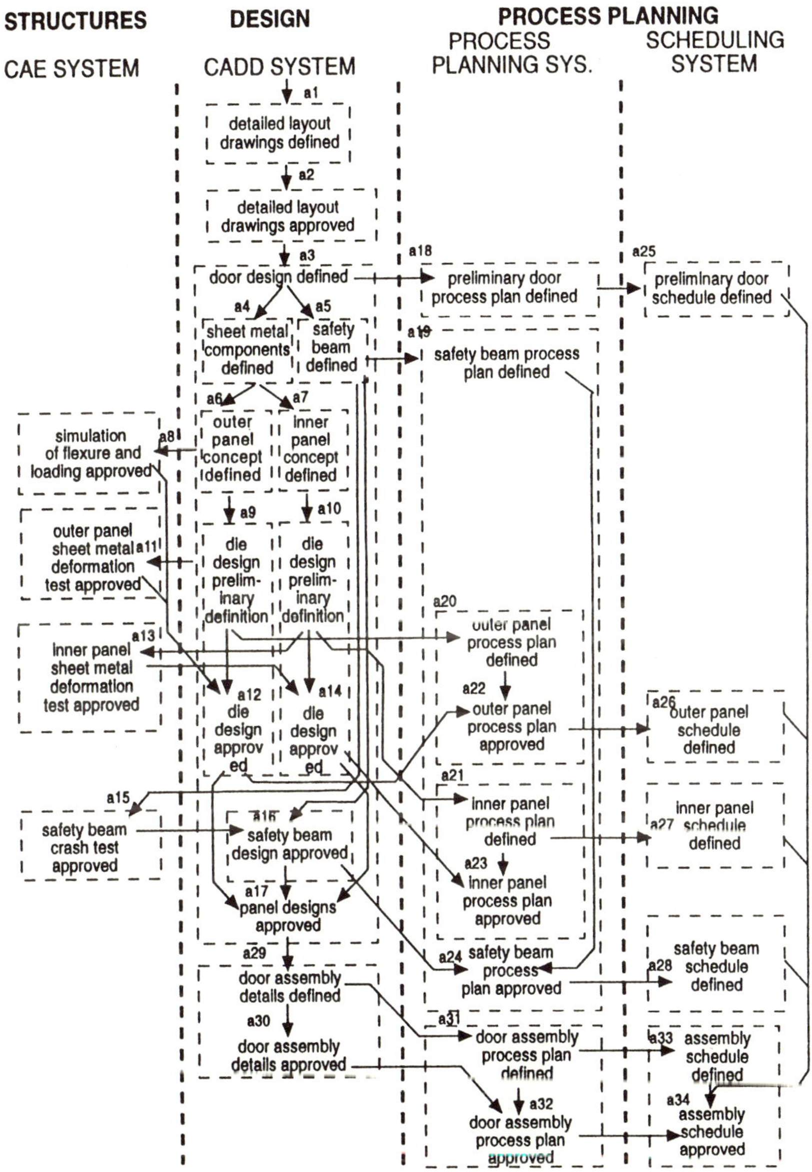

Figure 4: Task sequence for the example engineering problem after allowed state transformations have been revised in support of parallelism.

By specifying the logical conditions of use, a single document may go through several states before it is complete. The most significant example here is the general door specification dataset, which goes through five states:

C3. door design specified (on E2)
C4. sheet metal components defined (on E2)
C5. structural element defined (on E2)
C16. safety beam design approved (on E2)
C17. door panel design approved (on E2)

each of which denotes an incremental change that is significant for another document that uses the door panel concept design. By defining each of these incremental states, a higher level of parallelism is supported, which is evident by comparing Figure 4 with Figure 2. The result should be shorter overall design times.

The flow structure shown in Figure 4 involves many process choices. It incorporates, for example, the practice reported by Clark and Fujimoto how Japanese carmakers use preliminary designs of panels to initiate die designs[18]. This is shown in transactions a_9 and a_{10}. Process plans are also generated in parallel with the panel designs.

Iterations are managed within the EIS by the following process. Suppose that the outer panel design was initiated and both simulation of flexure and die design was initiated. Suppose that the flexure tests showed that the panel would vibrate under wind loads, requiring a re-design of the panel. When the panel is re-designed, that is transaction a_6, its completion is assessed according to eq.14, which identifies transaction a_9 as requiring review and possibly iteration.

4. Future Issues in Design Information Flow

In Section 2, the evolution of design was outlined, from a craftsman activity to a complex task involving the coordination of hundreds of experts. When design involved the flow of paper, the need for physical proximity was great. Now that means exist to distribute and manage design information electronically, the importance of physical proximity has greatly decreased. Other tools are under development to support design communication. These consider, for example, electronic group discussions and design reviews. These changes replace the need for product-based organizational structures[19].

---

18 Clark and Fujimoto, T. (September,1989), op.cit.

19 Stefik, et al, "Beyond the Chalkboard: Computer Support for Collaboration and Problem Solving in Meetings", Communic. of the ACM, 30:1, 1987, pp. 32-47.

EIS goes a long way to rationalize the flow of design information and to make it visible to study. For example, excessive iteration may be an indicator of contingencies in the design process. Unanticipated events in design processes may arise from a combination of errors in specification of the design activities to be executed; in the procedures used by the designers in executing the tasks; in communications between the different groups of designers involved in executing interrelated tasks; or in the procedures programmed into the tools used in executing the task. By studying the pattern of transactions, it may be possible to identify the source of such contingencies in the process. Study of the pattern of transactions might also allow for an identification of difficult information transfers and manual tasks for which new automated tools might be applied.

Evidence has been presented elsewhere of the importance of learning curves in fabrication[20]. It is suggested that similar learning curves might apply to iterated design operations, particularly where the sequence of design information processing is likely to change only in incremental ways. In these cases, it can be expected that the second iteration of the design will be more efficient than the first, the third quicker than the second and so forth. If so, then the reduction of cost in re-doing the design and the (probably) decreasing benefits of the iterations defines an interesting optimization problem regarding the optimum number of iterations. With decreasing costs, more iterations can be effective. Methods of promoting this type of "learning" across projects, such as by reuse of design information and design procedures, are also important new questions.

A related topic concerns the use of *experience* in design. One of its important applications is in making estimates of data:

- tasks that depend upon data generated in other tasks can be made parallel[21] by making estimates of the results of preceding tasks, allowing successor tasks to proceed. Later the estimate is verified, and if it was a good estimate, then the parallel task is already done. If the estimate was poor, then the task which used the estimate must be redone. Thus good estimates are very important in reducing design times.
- many analytic programs in design rely on convergence procedures. If the initial estimates are close to the final solution, the convergence will proceed with few iterations. If they are poor, then much more execution time for the analysis is required.

---

[20] Argote and D. Epple "Learning Curves in Manufacturing", Science, Volume 247, February 23, 1990, pp. 920-24.

[21] Clark and Fujimoto, T. (September,1989), op.cit.

It is clear that estimated data is an important component of expertise in engineering design. In the example of the automobile door design, the ability to generate good estimates for the specifications of sheet metal components makes it possible to initiate design of the dies before the component specs are approved. Poor estimates would lead to multiple die generation.

What is broadly suggested by this incomplete agenda for future research is that the information processing steps used to develop a design can be laid out in a precedence structure and the steps subjected to analyses, with regard to costs and efficiency of operations, in a manner similar to the steps in manufacturing. The operations are not mechanical ones, involving the transformation of materials, but rather information transformation processes that involve both multiple iterations and modification of flows as design proceeds.

At the level of a project manager, the EIS provides information at a level of granularity that was previously unavailable. It provides a mechanism for monitoring completion times of design steps, and to monitor the progress of the project in a manner which is largely innocuous to the organizational units. This mechanism overcomes many of the problems inherent in the accounting based methods developed in the 1930s, but which remain in widespread use today. By estimating the expected number of iterations for certain tasks, cost estimates for a task should be improved. By having access to detailed milestones within larger tasks, design costs can be better controlled.

It is important not to overstate the capabilities of EIS systems as they presently exist. EIS technology provides the ability to analyze the design process at the level of granularity of the document, which is smaller than was previously possible. Nevertheless, at this level of granularity, the parameters regarding sub-steps within an operation are not apparent. This imposes some limitations on the ability to predict time and cost resources within a design step. The shortcoming associated with the information granularity of the data set is also evident, for example, when a single change is made to a document, and all datasets dependent on the changed one are flagged as suspect. In reality, a change in one dataset may only affect the validity of specific data in other datasets. Work is currently underway that would transfer engineering data coming from diverse sources into a new generation of backend databases, with the potential ability to manage information flows of design data at the level of granularity of individual variables[22].

---

[22] Eastman, C.M.., A. Bond and S. Chase, "A formal approach for product model information", Research in Engineering Design, 2:2, 1991, pp. 65-80; Eastman, C. M., A. Bond and S. Chase, "Application and evaluation of an

5. Conclusion

Design often has been considered by management as a black box, which has a known production capability for known products, based on previous performance. Design schedules can be well defined for products similar to existing ones, but became poorer as deviations from previous designs grow. Schedules are based on previous resource expenditures, which are used to track design progress[23]. What we have shown here is that design operations can be decomposed in a manner similar to those of manufacturing, with individual design steps defined at the level of a document or dataset. The operations involved can be planned in a manner similar to manufacturing, but with important differences: iterations are not errors, but generally hill-climbing or convergence processes, and flows change during design, in response to unforeseen issues or opportunities. EISs provide a structure making the individual design operations apparent and analyzable. Design information management is expected to proceed from document level flow management to finer grains, dealing with states and then, as our understanding of design continues, to possibly the level of individual variables.

The decomposition of design tasks reviewed here complements work in cognitive psychology that is working to define design operations at the level of engineering problemsolving. At some point in the not distant future, these two lines of work may converge, allowing compatible models and estimation procedures to be used to plan and coordinate design activities.

---

engineering data model", Research in Engineering Design, 2:4, 1991, pp.185-208.

[23] Putnam, L.I. and A. Fitzsimmons, "Estimating Software Costs", Datamation, 3 part series, 25:9, pp. 89-98, 25:10, pp 171-178, 25:11, pp. 137-140.